MANAGING
THE RIVER
COMMONS

A VOLUME IN THE SERIES

Environmental History of the Northeast

Edited by
Anthony N. Penna and Richard W. Judd

MANAGING THE RIVER COMMONS

Fishing and New England's

Rural Economy

ERIK REARDON

University of Massachusetts Press
Amherst and Boston

Copyright © 2021 by University of Massachusetts Press
Printed in the United States of America

ISBN 978-1-62534-584-4 (paper); 585-1 (hardcover)

Designed by Sally Nichols
Set in Monotype Baskerville Pro
Printed and bound by Books International, Inc.

Cover design by Derek Thornton, Notch Design
Cover art by Thomas Cole. Detail from *View from Mount Holyoke, Northamption, Massachusetts,
after a Thunderstorm—The Oxbow*, 1836. Oil on canvas. Gift of Mrs. Russell Sage, 1908.
Metropolitan Museum of Art. CC0 1.0 Universal (CC0 1.0).

Library of Congress Cataloging-in-Publication Data
Names: Reardon, Erik, author.
Title: Managing the river commons : fishing and New England's rural economy
/ Erik Reardon.
Description: Amherst : University of Massachusetts Press, [2021] | Series:
Environmental history of the northeast | Includes bibliographical
references and index.
Identifiers: LCCN 2020053367 (print) | LCCN 2020053368 (ebook) | ISBN
9781625345844 (paperback) | ISBN 9781625345851 (hardcover) | ISBN
9781613768402 (ebook) | ISBN 9781613768419 (ebook)
Subjects: LCSH: Fisheries—New England. | Sustainable fisheries—New
England. | Fishes—Conservation—New England. | Conservation of natural
resources—New England.
Classification: LCC SH221.5.N4 R43 2021 (print) | LCC SH221.5.N4 (ebook)
| DDC 639.2/10974—dc23
LC record available at https://lccn.loc.gov/2020053367
LC ebook record available at https://lccn.loc.gov/2020053368

British Library Cataloguing-in-Publication Data
A catalog record for this book is available from the British Library.

Portions of chapter four were previously published as
"Fishing and the Rural Economy Farmer-Fishermen and the
Merrimack River, 1800–1846" in the *New England Quarterly* 89, no. 1
(March 2016): 54–83.

FOR M.W.

CONTENTS

PREFACE

I grew up a stone's throw from the Merrimack River close to the once-booming factory towns of Lawrence and Lowell, where repurposed mill buildings, canals, and large hydroelectric dams provide a glimpse into the region's industrial past. This postindustrial landscape inspired me to consider the complex relationship between environmental and economic change and sparked an interest in how rivers, as well as the aquatic life found therein, have been transformed by centuries of human activity. I began to wonder why the once-celebrated Atlantic salmon and other native species of fish have all but disappeared from their historic habitat and why environmental organizations are working so hard to bring them back. The prominence of river restoration in the contemporary environmental agenda also inspired me to think about how these efforts are connected to historical commitments to river conservation here in New England.

During my graduate work at the University of Maine, I was lucky enough to live in an old farmhouse on the banks of the Penobscot River in Orono. Not wanting to take this waterfront location for granted, I acquired an aluminum flat-bottom boat—ironically christened "The High Life"—and equipped it with a modest two-and-a-half-horsepower motor that just barely pushed the boat upriver against the current. I and my good friend Newt, the better fisherman of the two of us, spent the better part of two summers puttering up and down a small section of the river between two large dams, fishing for smallmouth bass. In June 2012, we approached an eddy near a series of rapids frequently visited by local bass fishermen. Directly to my right I saw a fish floating on the top of the water. It was so large that we were both initially puzzled as to what it could be. Then we saw the pink flesh exposed underneath its silvery skin and recognized the fish as an Atlantic salmon.

My first encounter with this species was altogether disappointing. It was not majestically leaping from the water in a trademark display of athleticism but most likely the victim of an eagle that swept down and attempted to pull the fish from the water. After months of fishing the river for bass, pickerel, and perch—species that occupy an ecological niche once filled by Atlantic salmon and other native fish—the experience showed me that this was a river transformed by dams and the introduction of invasive species. The U.S. Fish and Wildlife Service currently lists the Atlantic salmon under the Endangered Species Act, and although the Penobscot River still hosts the largest returning population of Atlantic salmon in North America, annual returns rarely exceed one thousand fish. This is a paltry figure considering the watershed was historically celebrated as the most productive salmon river in the northeast. Penobscot salmon were held in such high esteem that beginning in 1912, the first fish caught in the river each year was ceremoniously delivered to the president of the United States. It is possible that, in its heyday, the annual migration of Atlantic salmon to the Penobscot witnessed more than one hundred thousand fish returning to their natal streams.

Never having witnessed the famous salmon runs of the Pacific Coast, my only encounters with the Salmonidae family were in the form of a landlocked variety on the West Branch of the Penobscot River. For the uninitiated, native Atlantic salmon are truly a rare sight in the wild. This was not always the case. Atlantic salmon were just one of many migratory, or anadromous, species that once frequented New England's waters.

This book would not have been possible without the guidance and support of friends, family, and colleagues. I thank Richard Judd for introducing me to the field of environmental history. His scholarship has had a profound influence on my work and that of countless others within the discipline, and he has far exceeded what can be reasonably expected of any adviser in spending countless hours discussing this book and editing early drafts. His thoughtful consideration of my work has been invaluable and has undoubtedly helped me become a stronger writer and a better historian. Liam Riordan's early American writing seminar taught me a great deal about building a cohesive manuscript, and his suggestions for this book helped lead me down unexpected paths. Both Christopher Clark and Jeff Bolster were incredibly generous with their time in sharing their expertise as I began to work through source material. Matt McKenzie provided feedback that challenged me to sharpen my analysis and strive for greater clarity. The research staff at the Massachusetts, New Hampshire, Vermont, and Maine state archives worked diligently to identify relevant collections and fielded a seemingly never-ending series of questions. I also want to extend my heartfelt gratitude to my friends and family for helping me along this path, particularly my father, Kevin, to whom I owe my love of history. Finally, I thank my partner, Meghan Willis, for providing much-needed support and encouragement throughout the process of researching and writing this book. Her love for the natural world and passion for wilderness has inspired me to think deeply about pressing environmental issues. I simply could not have finished this project without her.

E.R.

Orono

February 2018

MANAGING
THE RIVER
COMMONS

INTRODUCTION

In 1857, Vermont governor Ryland Fletcher commissioned George Perkins Marsh to investigate the prospects for restoring the state's freshwater fisheries. Decades of overfishing, dam obstructions, and industrial pollution had reduced Vermont's native fish populations to near extinction. By the mid-nineteenth century, rivers throughout New England suffered from a similar combination of commercial impacts, leaving fishermen, politicians, and resource managers to wonder whether anything could be done to bring river fish back to their historic habitat. Marsh, the godfather of the conservation movement, enjoyed a distinguished political and diplomatic career; however, as Vermont's first fish commissioner, he expressed a particular zeal for confronting head-on the environmental problems of his day. In his report on the artificial propagation of fish, he lamented the destruction of wildlife, including "fish, and other aquatic animals," that often accompanied "advanced civilization and the increase and spread of a rural and industrial population." Having witnessed these declines firsthand at home and abroad, Marsh understood all too well that humanity possessed the unique capacity to disturb, disrupt, and ultimately undo the ecological connections that once supported healthy and productive forests, streams, and soils. Nevertheless, he remained convinced that "we may still do something to recover at least a share of the abundance which, in a more primitive state, the watery kingdom afforded."[1] Though Marsh would not live to see this ambitious goal made a reality, his hope for the eventual restoration of native migratory species has carried into the twenty-first century.

For New England and much of the Northeast, native species of anadromous fish belong to past generations. Anadromous fish migrate between rivers and the sea, and until the mid-nineteenth century they swarmed freshwater systems throughout the Northeast in astounding numbers. Schools of shad and alewives numbered in the millions on large main-stem rivers prior to the installation of industrial dams. While less abundant, Atlantic salmon were equally celebrated for their size, athleticism, and significant position within the regional diet. Together, these species not only furnished agricultural communities with an important source of nutrition but also offered a valued article of exchange within New England's rural economy. In his 1857 report, Marsh described how freshwater fisheries had once supplied Vermont farmers with "a very important contribution to the nutrition of a population with which the cultivated products of the soil were scarcely adequate to sustain."[2] Though agriculture stood at the foundation of economic life for households across the coastal lowlands and river valleys of southern New England, farmers throughout the region supplemented activity in the fields with hunting, gathering, timber harvesting, small manufactures, and the seasonal harvest of river fish. As such, legislation to regulate and protect river fisheries from exhaustion often promoted this particular resource as a "great public utility" that provided agricultural communities with "subsistence and support."[3]

With river fishing overwhelmingly regarded as a public good, this book argues that New England's farmer-fishermen sought to chart a course that would ensure sustainable river fisheries. They advocated for regulations designed to restrain commercial fishing operations, compelled local millers to open their dams during seasonal fish runs, defeated corporate proposals to erect industrial dams, and asserted a traditional vision of resource rights rooted in the culture of the commons. These efforts underscore an ethos of stewardship and conservation that ran through decades of petitions to state legislatures expressing deep concern for the future of this resource. Amid demographic and economic pressures, many remained convinced that river fisheries were worth saving.

Fish and fishing are central to New England's maritime heritage, yet as an open-access resource, fisheries also show an unfortunate history of decline, depletion, and ecological collapse. Alongside furs, timber, and moderately productive soils, abundant coastal and off-shore fish stocks underwrote European colonial prosperity in the Northeast, but with each new innovation in fishing technology, generations of European, American, and Canadian fleets pressed hard on Newfoundland's Grand Banks and the Gulf of Maine ecosystem. In *The Mortal Sea: Fishing the Atlantic in the Age of Sail*, W. Jeffrey Bolster notes that fishermen sounded the alarm of declining yields, but popular attitudes that "the sea was eternal and unchanging" overwhelmed voices that "raised disturbing questions about the perpetuity of the stocks on which they relied." Bolster writes that in the nineteenth century, "pressured by commercial capitalism and cornucopian fantasies, the northwest Atlantic's coastal ocean rapidly became an extension of Europe's diminished sea."[4] Serial depletions left an indelible mark on the northwest Atlantic as steam trawlers and purse seines swept schools of once commercially productive cod, halibut, mackerel, herring, menhaden, and the like from the Gulf of Maine as well as distant offshore banks.

Not nearly as intertwined with the region's economic fortunes or pursued on a scale that matched New England's offshore fleets, river fisheries nevertheless suffered from persistent commercial pressures that threatened sustainable returns. Starting in the 1750s, coastal and tidewater merchants along the Eastern Seaboard, from Maine to Connecticut, shipped preserved fish, including alewives, to sugar plantations in the West Indies to feed African slaves.[5] By the early nineteenth century, land shortages in the fertile, yet increasingly crowded, Connecticut River Valley pushed river fishermen to pursue commercial outlets for their catch, using boats and scoop nets to take thousands of fish from the water each day, which, according to John Cumbler, "brought the market deeper into the lives of the valley's residents, farmers, and fishermen." Around the same time, ships from southern

New England flocked to Maine's rugged coast to purchase barrels of pickled shad and smoked salmon for retail in eastern seaport markets. From 1842 to 1867, a single firm in Boston purchased nearly the entire annual harvest of shad from Maine's largest river systems.[6] Even as coastal, urban, and lucrative transatlantic markets drove these unsustainable practices, the legislative record echoed the fundamental principle that removing or obstructing a critical mass of fish, year after year, fed into a downward spiral that would prove difficult to reverse.

That commercial and, later, industrial capitalism transformed freshwater ecology and diminished inland fish stocks is beyond question, but according to Daniel Vickers—a leading historian of colonial economic culture and labor—market incentives alone do not explain recurring complaints of overfishing. For Vickers, rural communities, operating on the margins between commercial and subsistence production, contributed to resource decline by fishing at a level that exceeded household subsistence all the while "competing over the means to competency." The quest for economic independence, a "competency," within a competitive rural economy encouraged households to manipulate "the world around them to extract from it a surplus they could trade in various local and regional markets to answer their needs."[7] With country farmer-fishermen, in addition to commercial actors, pushing this resource beyond its limits, historians Bolster and Vickers paint a bleak picture of wanton abuse at sea, on the coast, and on the banks of the Connecticut, Merrimack, and Penobscot Rivers. On the surface, these interpretations align with well-established narratives of environmental destruction and ecological calamity that accompanied European colonization. William Cronon's *Changes in the Land* famously brought the environment to the forefront of colonization by arguing that European settlement caused fundamental shifts for the region's Indigenous communities and also New England's precontact ecology. Not long after the British solidified their place in this new landscape, they cleared forests at lightning speed, depleted soils, and reduced native wildlife to near extinction.[8] The exploitation of fisheries then seems a logical extension of British economic designs, funneling New World resources to transatlantic commodity markets. The sheer volume of inland fisheries regulations, saturating the legislative record from the colonial period to the mid-nineteenth century, points to similar patterns of Indigenous dispossession and European degradation, but such an overwhelming focus on narratives of uninterrupted depletion obscures popular commitments to economic independence that fostered sensitivity for the resources supporting New England's early agricultural communities.

This book argues that commitments to sustainable fisheries accompanied observations of resource decline, and for a time, an environmental sensibility, rooted in the values, norms, and traditions of New England's rural economy, served to counteract commercial intrusion into the traditional resource base. River fisheries occupied a central position within the region's agrarian economy and corresponding rural exchange networks. As a result, farmer-fishermen guarded against commercial and industrial forces that could potentially jeopardize rural subsistence and household independence. Experiential knowledge of the timing and variability of anadromous migrations, the location of spawning habitat within a given watershed, and the necessity of sufficient stock to complete the life cycle and reproduce the next generation revealed to petitioners the fragile conditions that held this resource together. As these communities witnessed unsustainable and destructive impacts associated with commercial fishing and the proliferation of mill dams, they petitioned lawmakers to intervene with regulations derived from their own experiences of local environments. Historian Ruth Bogin argues that petitions from the revolutionary era "give us the voice of people who seldom if ever proclaim their social goals and political opinions in other written forms."[9] The petitions featured in this book reveal farmer-fishermen working to preserve an environmental heritage that buttressed rural social structures—encompassing productive soils, meadows, forests, streams, and main-stem rivers—and wary of the commercial turn that would place new demands on the natural world. More than a simple plea to save the fish, petitioners strove to defend communal rights to a traditional resource.

From the colonial period to the mid-nineteenth century, farmer-fishermen partnered with state legislators to extract significant concessions from commercial fishers within the tidewater and inland spawning grounds. They placed limits on the duration of the seasonal weir fishery within coastal estuaries and the use of large nets and seines above the head of tide, all the way to upland headwaters. And though some historians posit that this resource was doomed regardless of industrial dams, the history of river fishing in New England points to a long process of cooperation and experimentation between legislators and fishers and a deep commitment to identifying the correct balance of regulations that would yield the desired result: maintaining or restoring sustainable fish populations. Industrial dams abruptly terminated this search for the appropriate mechanisms to deter and punish overfishing. Perhaps a return to levels of abundance described by early European settlers of North American was unattainable for petitioners, but a functioning equilibrium was not out of the question. The ways in which these farmer-fishermen responded to threats against their livelihood highlight the integral nature of fish within the rural economic landscape, an ethos of environmental stewardship in the New England countryside, and connected efforts to sustainably manage a shared resource before the advent of industrialization.[10]

The economic culture of the New England countryside may help explain why so many fishermen felt empowered to determine appropriate conduct with respect to the river commons and other critical natural resources. With widespread landownership and soils often ill-suited for staple production, rural households primarily busied themselves providing for their basic needs. This was especially true for colonists who settled the inland and upland boundaries of the New England frontier during the eighteenth century. In the absence of navigable rivers, poor roads rendered market production a moot point.[11] Under these circumstances, farmers participated most enthusiastically in local networks of community trade, bartering goods and services with one another to acquire what they could not produce on their own and engaging regional markets with whatever surplus might remain after their needs had been met.[12] Through a combination of family labor and reciprocal community exchange, these farmer-fishermen embraced seasonal, multi-occupational pathways to secure a comfortable subsistence. In *The Roots of Rural Capitalism*, Christopher Clark notes that prior to 1830, farmers in western Massachusetts devoted most of their energies to "the independent household economy in which the use of wage labor and production for distant markets were severely curtailed."[13] Households might also pursue hybrid strategies, blending subsistence and market production, but according to James A. Henretta, across the northern colonies "men and women were enmeshed in a web of social relationships and cultural expectations that inhibited the free play of market forces." Capitalist pressures to specialize production or increase yields took a back seat to lineal values that privileged the "the yearly subsistence and the long-run financial security of the family unit."[14] A valued and predictable source of seasonal nutrition, river fish contributed directly to household subsistence and, in conjunction with farming and a whole host of natural resources, economic independence in the countryside. Removing one component of the resource base was likely to create a ripple effect that could ultimately disrupt the integrity of the agrarian economy for future generations. One might, as Vickers suggests, actively pursue surplus resources to acquire additional lands so as to divide an inheritance between several sons,[15] but yeomen farmers also worked to maintain the extensive resource base that held the rural economy together. To pass down a sustainable agrarian legacy to the next generation, subsistence farmers embraced a long view in balancing the interdependent elements of the farm ecosystem: livestock, fields, orchards, gardens, pastures, meadows, and woodlots.[16] Decades of petitions and rural account books demonstrate that this ethos of stewardship encompassed the entirety of the agrarian resource base, including the river commons.

With such strong generational values embedded within the rural consciousness, perhaps it is no surprise that historians have identified New England as a staging ground for experimentation with sustainable environmental practices. As Richard Judd argues in *Common Lands, Common People*, early expressions of conservation thought emerged from popular folk traditions of land and resource management in northern New England. Blessed with prolific forests, fresh water, and coastal resources, rural communities interpreted rights and responsibilities to local environments through the lens of democratic access and common stewardship.[17] Pushing the chronology back to the colonial era, Brian Donahue found generations of farmers in Concord, Massachusetts, operating within the ecological limits of the local environment. Colonial farmers certainly transformed the world around them in order to re-create familiar European agricultural regimes—at least as much as the climate, geography, and soil composition would allow—but commitments to common stewardship of the land also shaped the character of this transplanted agrarian culture.[18] While the first several generations of colonists focused on survival and security and thought little for conserving nature, by the middle of the eighteenth century demographic expansion, coupled with the limited availability of arable land, brought resource conservation into the realm of practical self-interest.[19] In a similar vein, farmer-fishermen entered the river commons with a distinct purpose: to use this resource to provide for the material needs of the household. They invested significant time and energy to acquire their fair share of the seasonal fish runs, which undoubtedly had some impact on the broader freshwater ecology. The Market Revolution ratcheted up the pressure to decidedly unsustainable levels. Farmer-fishermen believed they stood on the correct side of the boundary between reasonable use for subsistence and local trade and commercial abuse.

This book begins with North America's first inhabitants, the Indigenous peoples of the Northeast, who sustained their communities through an elaborate regime of hunting, gathering, planting, and fishing. Known for their seasonal mobility, New England's Algonquian-speaking peoples fished the coastline, tidewater, and upriver environments for countless generations prior to European contact. Satellite fishing camps in southern New England and intertribal gathering sites along inland waterfalls appear prominently within both the archaeological record and early European accounts of Algonquian subsistence culture. Early European explorers marveled at the seemingly limitless abundance of coastal and freshwater fish and expressed admiration for the skill with which Indigenous fishermen harvested this seasonal resource. Seasonal fishing represented a point of common interaction and then conflict as British settlers quickly capitalized on Indigenous knowledge of productive fishing locales and built brush weirs along the coastline of southern New England. By 1650, localized depletions within these coastal and estuarine ecosystems forced colonists to confront the possibility that New World resources were not as inexhaustible as early chronicles would have them believe.[20] To combat these unwelcome trends, Euro-American communities drew on recent memories of resource decline in Europe and incorporated the kind of top-down administrative efforts often seen in Great Britain to prevent the collapse of coastal and inland fish stocks. Further upriver, colonists also borrowed traditions from English common law that for centuries had prevented mill owners from blocking the upstream migrations of anadromous fish.[21] New England's rural fishing communities, looking forward while also applying the lessons of the past, absorbed these legal and environmental precedents to pull river fisheries from the brink of collapse.

The river commons ran deep within the complex web of seasonal production that constituted New England's preindustrial, rural economy, setting the conditions for an agrarian conservation ethic that privileged the long-term survival of subsistence fishing traditions. A composite system of mixed husbandry, gendered division of household labor, and durable patriarchal authority, New England's rural economy persisted in its core structure for two hundred years, from the beginning of colonial settlement and well into the nineteenth century.[22] In the middle of the seventeenth century, colonists began to push inland from the coastal lowlands, following the region's major river systems and relying on seasonal fishing as a vital foothold in the frontier forage economy. Both inland

and coastal farm communities incorporated seasonal river fishing into their pursuit of a comfortable subsistence. Account books, diaries, and mid-nineteenth-century oral histories provide a window into the working lives of farmer-fishermen, the central position of fish in community exchange networks, and the formation of informal cooperative fishing arrangements between community members. After the American Revolution, market structures increasingly drew coastal fishermen into the world of Atlantic commerce, which in turn motivated the types of overfishing inland farmer-fishermen wished to restrain. By the turn of the nineteenth century commercial fishermen deployed weirs in the tidewater and seines and other large nets above the head of tide to trap large numbers of fish during their upstream migrations. A resource uniquely vulnerable to over-exploitation, anadromous fish travel in large schools over short periods of time, and commercially oriented fishers often took advantage of this feature to harvest as many as their equipment would allow.[23] Positioned below a waterfall, inside a narrow channel, or near the mouth of a brook or stream where fish entered to spawn, a few resourceful individuals deployed specialized gear and methods to monopolize entire runs of fish. For decades, farmer-fishermen, commercial harvesters, local millers, and, later, corporate actors struggled to sort out competing interests in freshwater and river fisheries. For the farmer-fishermen who relied on this resource for personal consumption, rural exchange, and household independence, overfishing and industrialization threatened not only a valued natural resource but also their agrarian economic culture.

Campaigns to preserve river fisheries emanated from the grassroots where fishermen felt they possessed the requisite environmental knowledge necessary to prevent the collapse of this resource. With the proliferation of small mill dams, the opportunistic and monopolistic approach of a bourgeoning class of commercial fishermen, and the specter of industrialization looming, communities of country fishers petitioned state legislators for restrictions to preserve inland fish populations. Time and again, New Englanders identified river fish as an important resource worthy of protection.

For this reason, it is important to recognize that the decline of river fisheries and the transition to a manufacturing economy were far from predetermined. In the countryside, commercial and industrial actors ran up against deep commitments to economic independence and sustainability. In fact, in the case of river fisheries, this agrarian conservation ethic was so formidable that a coalition of rural interests forestalled an industrial takeover of the Merrimack River in the early nineteenth century, a historical moment historians Morton Horwitz and Theodore Steinberg associate with manufacturing firms consolidating the necessary economic and political influence to displace traditional water rights and erect large industrial dams.[24] Rural coalitions would not have worked to defeat corporate development initiatives in the early nineteenth century if they felt the resource itself was beyond salvation. They continued to oppose dam construction as late as the 1830s in central Maine, and even after manufacturers installed large dams, fishermen agitated for technical solutions to recover some measure of what had been lost. Taken together, this agrarian conservation ethos represented a potent legislative, regulatory, and ideological force from the colonial era until the mid-nineteenth century—a counterbalance to the rising tide of commercial and industrial power. Recognizing the eventual industrial and commercial takeover of the region's largest and, therefore, most profitable river systems, this book highlights moments in which rural, agrarian communities resisted commercial overexploitation and industrial uses of rives, instead emphasizing the contingent nature of environmental degradation and industrial conquest in the early national Northeast.

The conflicts over access and fair distribution of resources that inform this narrative have historically, and in many ways continue to, undercut efforts to sustainably manage natural resources. Garrett Hardin famously wrote in the "Tragedy of the Commons," that individuals inevitably overwhelm shared resources and sacrifice communal obligations for economic rationality and short-term self-interest. Nobel Prize–winning economist Elinor Ostrom defines common-pool resources (CPRs) as "natural resource systems used by multiple individuals." She further clarifies that "common-pool resources generate finite quantities of resource units and one person's use subtracts from the quantity of resource units available to others." The difficulty of excluding unsanctioned or profligate use also makes CPRs uniquely vulnerable to exhaustion.[25] The challenges of establishing and enforcing boundaries as well as, in some cases, excluding bad actors can complicate the sustainable management of any common-pool resource, which was certainly the case for New England's preindustrial river fisheries. Because river fisheries were largely open to the public, depletion was always a real possibility. Yet, over the past several decades, a handful of scholars have theorized that resource rights can operate sustainably at the local level. Pioneered by Elinor Ostrom, this trend toward decentralized conservation and regulatory rights emphasizes that those closest to the resource itself can manage for outcomes that balance the interests of stakeholders and larger concerns for sustainability.[26] Whether communal mountain meadows in Switzerland or irrigations systems in Spain, Ostrom and others show that stakeholders have drawn on shared values to craft complex and sophisticated sets of rules, norms, and institutions—over the course of generations—to effectively govern local resources. As Bonnie J. McCay and Svein Jentoft explain, "Community exists, it counts, and it shapes the nature and outcomes of commons problems." For James Acheson and Jack Knight, decades of stable landings in Maine's lobster fishery support the conclusion that "environmental disaster is not inevitable . . . for in many cases those who are dependent on resources, and their governments, have acted to generate rules to manage those resources at sustainable levels."[27]

In making the case that those directly invested in the survival of shared resources can develop responsible management systems, Ostrom identified sociocultural characteristics that set the conditions for sustainable use. Typically, she pointed out, stakeholders belong to relatively homogeneous communities, depend on the resource for their livelihoods, and share a common understanding of how the resource operates regarding use and extraction.[28] The ancestors of culturally homogeneous and communal transatlantic migrations, New England's farmer-fishermen shared a common Anglo-American heritage, relied on seasonal river fishing as a fundamental component of their agrarian subsistence, and communicated nuanced ecological understandings associated with fish migrations, spawning habitat, and stock replacement. Until the appearance of industrial dams, these qualities would contribute to the emergence and maintenance of a subsistence fishing culture and the consensus that only regulatory intervention could preserve this resource for future generations.

This book reaches back to a historical moment prior to the industrialization of the region's inland waters when communities of farmer-fishermen pushed back against disruptions in the river commons. They worked to prevent resource exhaustion, ensure sustainable returns, and implement a conservation vision framed by principles of equity and restraint. At the turn of the nineteenth century, New England's farmer-fishermen operated within a framework of community consensus and, according to an intimate shared knowledge of the fishery itself, engaged grassroots networks of like-minded farmer-fishermen that feared for the long-term viability of this resource.

Over the past few decades, river restoration has emerged as a core issue within the environmental movement. The conservation organization American Rivers proclaimed 2011 "the year of the river," as a result of dam removal projects and the restoration of habitat for native species. Environmental organizations and Native American nations have stood on the front lines of this conservation agenda, pushing state and federal agencies to reconsider the ecological impacts of dams. Their hard work has produced visible results. In Maine, dam removal projects on the Kennebec and the Penobscot Rivers demonstrate that native anadromous species are incredibly resilient and poised to reclaim their place within these ecosystems. These projects also show that there is good reason for all the public attention devoted to river conservation. River restoration projects connect to powerful memories of environments ingrained in regional culture. They offer hope that the conditions that once supported healthy rivers are once again within reach.

Given the attention currently placed on the restoration of river habitats and anadromous fish runs, this book offers a deeper understanding of the forces that shaped public attitudes and material relationships with rivers and inland fisheries. The current depleted state of the region's major watersheds indicates that some historical context could prove instructive for these restoration programs. New Englanders may have long forgotten the tremendous biodiversity these preindustrial rivers supported, perhaps the names of anadromous fish are unrecognizable to many riparian communities, but stories that document the overwhelming commitment to preserving this biodiversity can inspire a broader public awareness of the benefits that flow from healthy river systems.

CHAPTER ONE

CHANNELS AND FLOWS

In May 1635, colonists from Massachusetts Bay pressed Governor John Hayes, Deputy Governor John Winthrop, and the Court of Assistants, also known as "magistrates," to create a formal legal code to replace the company charter granted by King Charles I. As the colonial population swelled during the great Puritan migration, John Winthrop noted that "the people had long desired a body of laws, and thought their condition very unsafe, while so much power rested in the discretion of magistrates."[1] To limit the authority of the magistrates and open more space for representative government, the Reverend Nathaniel Ward of Ipswich drafted a provisional compact that outlined the distinct rights, liberties, and responsibilities of men, women, children, and servants within the colony.[2] Towns across Massachusetts Bay debated, revised, and amended Ward's proposal, and in 1641, the Massachusetts General Court—the seat of legislative and judicial authority—accepted *The Body of Liberties* as the first European legal code in New England. Informed by English common law and Puritan theology, this collection of ninety-eight articles formed the basis for criminal and civil law, which included protections for individual and communal rights to the land and water within the boundaries of the colony. Section sixteen defined public rights to coastal and inland waters, stipulating that "every inhabitant that is a householder shall have free fishing and fowling in any great pond and bays, coves and rivers, so far as the sea ebbs and flows within the precincts of the town where they dwell, unless the free men of the same town or the General Court have otherwise appropriated them." Above the head of tide, town officials and private landowners determined access to rivers and smaller tributaries, but in 1649 included the exception that "great ponds lying in common . . . within the bounds of some town, it shall be free for any man to fish and fowl there." The Crown granted the Massachusetts Bay Company exclusive authority over coastal waters and large navigable rivers, within the limits of the corporate charter, with the expectation that colonial leaders held these waters in trust for public use.[3] The freemen of the colony, male church members with voting rights, exercised this authority to establish public fishing rights within coastal and inland water and determined rules for entry and exclusion. *The Body of Liberties* thus adopted legal precedent that gave form to a river commons: a resource open to heads of households within defined, but fluid, colonial boundaries. This declaration of "free fishing" anticipated a rural economic culture that privileged river fishing within the seasonal round of colonial subsistence: livelihoods rooted in the forests, fields, meadows, and waters of the New England countryside.

From low-lying coastal estuaries to steep upland channels, New England's inland waters encompassed a spectrum of river types and ecologies that once sustained a diverse array of aquatic life. As with most watersheds along the Appalachian corridor, rivers begin as small but powerful mountain streams, narrow and straight as they descend through steep-sided canyons. From New Hampshire's White Mountains, famous for scenic alpine vistas and harsh, unpredictable weather, the headwaters of the Merrimack, Androscoggin, and Saco Rivers flow into wide, U-shaped valleys on their way to the Gulf of Maine. The Connecticut River, the region's largest watershed, begins north of the mountains with a series of four lakes near the border with Quebec. Also fed by underground springs, swift-moving, shaded mountain streams offer prime habitat for cold-water species like brook trout. As these streams exit mountain elevations they join with a network of wetlands, lakes, and tributaries to form more extensive channels where larger trout mingled with pike and pickerel.[4] Leaving the

FIGURE 1. "The New England Landscape," in William F. Robinson, *Abandoned New England: Its Hidden Ruins and Where to Find Them* (Boston: New York Graphic Society, 1976), 2.

foothills of the north country, these rivers enter broad valleys and meander past forests, rolling hills, and meadows. Along the banks of these large downriver corridors, Indigenous peoples, and later European settlers, tilled some of the most productive soils in the northeast. Rivers transport and deposit sediment, minerals, and nutrients from further upstream that over time create fertile alluvial soils. Indigenous and British colonial riparian communities also harvested the seasonal bounty of the river itself, particularly during migrations of anadromous species.

Diadromous fish migrate between freshwater and saltwater habitats. Referring to the specific direction of their migration, anadromous fish are spawned in freshwater, spend the majority of their lives in the ocean, and then return to the very same freshwater habitat to complete their life cycle. Catadromous species live in freshwater but migrate to the sea to spawn. Over a dozen species of anadromous fish once called the region home, but only one catadromous species, the American eel, is native to the region. To survive such remarkably different aquatic environments, diadromous fish evolved a unique biological mechanism to regulate osmotic pressure—the concentration of water and salts in body fluids. Biologists call these automatic physiological adjustments made between fresh- and saltwater habitats "osmoregulation." This evolutionary adaptation facilitates diadromous migrations, which in turn allow these species to take advantage of forage opportunities across both freshwater and marine environments. Marine productivity, particularly in the Gulf of Maine, far exceeds that of freshwater habitats, and for this reason biologists theorize that "food availability is an important factor in explaining not only where diadromous fishes occur, but also why fish migrate across the ocean-freshwater boundary, as well as their direction of movement."[5] Though fisheries biologists understand why these species migrate between freshwater and the sea, the question of how anadromous species such as Atlantic salmon traverse vast distances across the open ocean and then return to the exact place of their birth remains something of a mystery.

The return of anadromous fish to their natal waters is one of the most remarkable feats of the animal kingdom. After spending the majority of their lives in the sea, such anadromous fish as Atlantic salmon, shad, and alewives navigate back to the waters where they were born to spawn the next generation. Fisheries biologists disagree as to exactly how these species are able to "home" to the streams and ponds of their birth, but it may have something to do with chemical imprints that guide fish back to natal waters. Nevertheless, the return of anadromous fish provides a boon of marine-derived nutrients to the benefit of other freshwater species and the overall river ecology. Even though the transfer of nutrients is otherwise one-sided, as freshwater brings nutrients to the sea, the anadromous life cycle brings back a small measure of nitrogen and phosphorus as thousands, and in some cases even millions, of fish enter coastal estuaries and begin to swim upstream.[6] Historically, New England's rivers hosted as many as eleven diadromous species, including rainbow smelt, alewives, blueback herring, American shad, striped bass, sea lamprey, shortnose sturgeon, Atlantic sturgeon, Atlantic tomcod, Atlantic salmon, and American eels. Prior to the installation of large industrial dams in the nineteenth century, these species contributed to not only rich, productive riparian environments but also the material welfare of the region's Algonquian-speaking Indigenous peoples and, by the seventeenth century, British colonial settlers. Three species stand out in the archaeological and historical record for their abundance and prominent position within both Indigenous and Euro-American cultures.

Alewives, a species of river herring and one of the most abundant food fish for Indigenous and colonial communities near the tidewater, were once found along the Atlantic coast from Florida to Maine.[7] Beginning sometime around April, alewives begin their journey from the coast to inland spawning grounds, typically calm waters such as small lakes and ponds close to the estuary. Because alewives tend to spawn near the tidal reaches of a river, they offered a predictable source of nutrition for agricultural communities along these well-established migration channels. Both Algonquian peoples and colonial settlers constructed weirs in tidal waters, traditionally made of stone, brush, wooden stakes, and netting, to obstruct and trap fish in a series of small enclosures as they swam upstream. Further upriver, fishermen used seines, large nets extended vertically from the surface to the river bottom, weighted with sinkers, paid out into water with a boat, and then dragged by a company of men back toward the shore.[8] Native Americans and British colonists annually caught large quantities of alewives and consumed the fish fresh shortly thereafter or smoked them on large racks to preserve a portion of the catch for future use. Around the turn of the nineteenth century, commercial fishers sent hundreds of barrels of salted fish each year to merchants for sale in urban markets.[9] Though fishermen mostly pursued alewives to supplement the household diet and engage with community trade networks, they also supplied bait for New England's offshore cod fleets. The alewife was one of the most abundant and sought-after anadromous species, and records indicate that alewife runs stayed relatively strong well into the nineteenth century. Raymond McFarland estimated that New England fishermen collectively landed nearly ten million alewives in 1880 alone.[10] With spring alewife runs sometimes numbering in the millions, both Indigenous and colonial settler communities celebrated the beginning of the seasonal alewife run as a symbol of the coming spring season. To this day, locals and tourists alike gather to witness half a million alewives migrate up a small stream in Damariscotta Mills, Maine.[11]

Colonial and early national fishermen expended the most energy and resources during the migration of American shad, the largest species of the herring family. From April to June, shad journey back to freshwater. Their migrations can extend hundreds of miles into the interior, where they remain in freshwater for several months before some, especially those within the northern latitudes of their historic range, swim back to the ocean to repeat their spawning migrations over the next several years.[12] Before the construction of large dams on the Merrimack River, shad traveled as far as north as Lake Winnipesaukee, an astounding 125 miles from the coast, allowing inhabitants of the entire Merrimack River valley an opportunity to take advantage of this seasonal resource. As the shad move deep into the interior they pass into smaller tributaries, seeking larger and warmer bodies of water to spawn. Fishermen took shad with weirs, seines, and drift nets in lower reaches of a river but also used boats, seines, and dip nets close to waterfalls and other natural obstacles. New England's rural communities came to depend on the annual shad runs as an important article of their household subsistence, but early nineteenth-century commercial shad fishermen put significant pressure on this particular species. The Connecticut River, home to what was once one of the most abundant shad fisheries in New England, increasingly attracted a class of full-time commercial fishermen, particularly within the estuary. Beginning around 1849, fishermen built pound nets, a fixed equipment similar to weirs, that crowded the mouth of the river near Long Island Sound and hijacked entire schools of shad. Between 1866 and 1871, a single pound net in Westbrook, Connecticut, captured over ten thousand shad each year.[13] Juvenile shad also historically served an important function within the marine food chain, luring deep-sea fish such as haddock and cod to coastal waters. The downriver migrations of juvenile rainbow smelt, alewives, and shad toward coastal waters connected freshwater and marine ecosystems and contributed to productive inshore fisheries.[14]

The Atlantic salmon's habitat once stretched from as far north as Canada's Atlantic provinces and then south to the Housatonic River in Connecticut. One of the longest anadromous migrations in terms of duration, Atlantic salmon leave the ocean at various points throughout the spring, summer, and early fall.[15] Famous for their feats of athleticism, salmon leap incredible distances over seemingly insurmountable obstacles to reach upland spawning grounds. Appropriately, the latter portion of the Latin name, *Salmo salar*, translates as "the leaper." Though the Connecticut and the Merrimack Rivers once hosted large populations of this anadromous species, the state of Maine laid claim to the most abundant Atlantic salmon fisheries in the northeast. The Indigenous Wabanaki peoples fished deep pools from their canoes with torches and spears. When the soils failed to accommodate productive farms, rural fishermen on the coast of Maine depended on inshore and river fishing to eke out a living. This species also attracted commercial pressure, with fishermen employing both mobile and fixed gear, including drift nets, pound nets, weirs, and seines. Starting in the 1820s, merchants sent schooners up the Atlantic coast to the Penobscot River to purchase barrels of cured salmon for markets in Boston and New York.[16] In addition to impacts from commercial overfishing, salmon suffered the most from callous obstruction of upland spawning habitat.

Unfortunately, because Atlantic salmon require access to the entire watershed to complete their life cycle and spawn the next generation, this species was uniquely vulnerable to human interference. According to fisheries biologists, environmental conditions must permit "unobstructed access between freshwater, estuarine, and marine environments. In these variable ecosystems, an intricate set of events is required for salmon to successfully complete their life cycle."[17] Before the construction of large industrial dams in the nineteenth century, salmon migrations extended deep into the upland headwaters of the region's largest river systems, which afforded rural farmer-fishermen opportunities to access these seasonal fish runs. Prior to the 1820s, which saw the factory town of Lowell, Massachusetts, emerge as the paradigm of industrial manufacturing, both salmon and shad ascended the Merrimack River to Franklin, New Hampshire, at which point the salmon followed the Pemigewasset River toward the cold streams of the White Mountains.[18] Atlantic salmon thrive in cold-water environments. The colder the water, the greater the concentration of dissolved oxygen, which is necessary to sustain anadromous species like Atlantic salmon. Within these mountain streams, between October and November, female salmon bury their eggs in gravel river bottoms, which offer some protection against predators; the eggs hatch the following spring.[19] These young salmon, called "parr," typically remain in headwater streams for as long as six years until they undergo a host of physiological, hormonal, and behavioral changes that prepare them for their ocean journey. Now classified as smolts, they finally resemble the silvery adults they will soon join in the sea. After a two-year journey across thousands of miles of open ocean—intermingling with other salmon from Canada, Greenland, Iceland, and Europe—Atlantic salmon return from feeding grounds in the Labrador Sea as adults weighing upward of ten pounds. Unlike their Pacific relatives, Atlantic salmon may repeat this cycle more than once.[20] Once abundant in large and small river systems across the Northeast, the salmon's historic habitat has shrunk to only a few rivers in Maine, and even in those remaining few the federal government lists them as an endangered species. For this reason, mid-nineteenth-century state fish commissions and, more recently, environmental organizations have expended significant time and resources to restore Atlantic salmon to their native waters.

Every spring and summer, New England's rivers were once filled with incredible numbers of anadromous fish, furnishing Native American communities with a predictable food source and contributing to the overall health of freshwater ecosystems. Long before European settlement, Native Americans moved seasonally across the Northeast woodlands, relying on rivers for transportation, communication, trade, and sustenance. They fashioned distinct traditions, which survive to this day, that speak to the cultural and material significance of the rivers they once called home and the anadromous fish that helped provision their communities. With the arrival of European colonists to the coast of New England, the ancestral homelands of diverse Indigenous communities, anadromous fish would come to occupy a prominent space in both Native American subsistence and Euro-American agricultural economies. Taken with the abundance of New World fisheries, colonial authorities in Massachusetts Bay established familiar common resource regimes that allowed for free and open fishing along the coast and tidewater, soon pressuring this resource to the point of collapse. This myth of unlimited abundance did not last long as the first concerns for the survival of inland fish stocks were not voiced alongside the installation of industrial dams in the nineteenth century but early in the colonial period.

CHAPTER TWO

COLONIAL ENCOUNTERS

Indigenous and European
River Traditions

In 1991, state officials conducted a drawdown of Sebasticook Lake in Newport, Maine. Originally designed to address problems of pollution and eutrophication, as the lake began to recede over one hundred wooden stakes rose out of the water at an outlet of the East Branch of the Sebasticook River on the northern end of the lake. Archaeologists called to the scene from the University of Maine at Farmington quickly understood what they were dealing with and identified the remains of an ancient fish weir. An efficient fishing technology that reaches back to prehistory, weirs consist of materials arranged in a river or stream that create a fence-like obstacle to guide and trap fish passing upriver. Driven deep into the bed of the river and well preserved beneath the water, the wooden stakes that emerged from the Sebasticook River, along with stone artifacts that accompanied the structure, brought new understandings of Native American relationships with freshwater ecosystems thousands of years before European colonization.

Radiocarbon testing revealed that the oldest wooden stake dates to an astounding 5080 BP, which would perhaps constitute the oldest weir complex in North America. With additional stakes dated to 1760 BP, archaeologists suggest that Indigenous communities maintained multiple weirs at this site across thousands of years. This must have been a very productive location for river fishing as such a complex structure, reassembled and maintained over thousands of years, would have required significant collective effort.[1] Investigators also identified a possible net-sinking stone along with the remnants of a birch bark container, artifacts commonly associated with seasonal fishing camps. Though much remains to be discovered about the position of freshwater ecology within Indigenous subsistence cultures, the Sebasticook site confirms that people have been fishing New England rivers for a very long time.

Part of the Wabanaki ancestral homeland, the Sebasticook flows southwest to the Kennebec River, which continues south to Merrymeeting Bay before finally emptying into the Gulf of Maine. As part of the larger Kennebec River basin, linking both marine and inland freshwater ecosystems, Indigenous communities that maintained the Sebasticook weir site, whether five thousand years ago or at the time of colonization, inhabited prime migratory fish habitat. Seventy-five miles above the tidal waters of Merrymeeting Bay, schools of salmon, shad, and eels could have traveled above the estuary all the way to the Sebasticook River. Archaeologists have found bone fragments of Atlantic salmon and shad that further confirm the presence of migratory species thousands of years ago on both the Kennebec and Penobscot watersheds.[2] The seasonal arrival of salmon, shad, alewives, and other anadromous species offered a relatively predictable resource that would have encouraged intensive exploitation during the spring and summer, and according to evidence gathered at the Sebasticook weir complex, Wabanaki peoples occupied these prime fishing grounds for countless generations. Algonquian-speaking people who inhabited ancestral lands that now comprise New England relied on coastal and inland fishing as part of their seasonal subsistence and apparently pursued these resources without causing significant disruptions to the larger marine or freshwater ecology. Early European chronicles invariably characterized this "new world" as a veritable Garden of Eden. Colonists were clearly taken with the abundance and diversity of plants and animals found in the fields, forests, shores, and inland waters of the Northeast. Writing in the

1630s, English colonist Thomas Morton could scarcely believe the "multitudes of fish, both in the fresh waters and also on the coast, that the like hath not elsewhere been discovered by any traveler." Near his home at the trading colony of Merrymount, present-day Quincy, Massachusetts, Morton supposed that "one should not throw a stone in the water but that hee [sic] should hit a fish."[3] These early colonial descriptions, combined with evidence from precolonial Indigenous sites, paint a picture of productive freshwater ecosystems marked by seasonal abundance and incredible levels of biodiversity.

Early in the colonial period, seasonal river fishing represented an important component of Indigenous and European subsistence and would come to mark a point of common interaction. Europeans admired the skill and expertise of Indian fishermen. For the first few decades of settlement, colonists appropriated this Indigenous knowledge, accumulated over thousands of years, and used it to gain a foothold in the unfamiliar environment. Not long after, competition over land and diminishing fish stocks strained relations between these two cultures as colonial settlements continued to encroach on river systems that once comprised the Algonquian homeland. The effects of this colonial dispossession appeared soon thereafter. Less than a century after British colonization in New England, the Province of Massachusetts Bay, dismayed by substantial declines in seasonal fish stocks, instituted restrictions to protect the resource from collapse.[4] Fast-growing colonial populations stressed their local environments, which forced fishermen and colonial authorities to confront the unfortunate patterns of environmental decline that accompanied settler colonialism. Something would have to change to bring the fish back. Euro-Americans would spend the next century trying to identify the most effective way to accomplish that goal.

The Native Northeast

The seasonal rhythms of New England's climate and ecology profoundly shaped the character of Native American lifeways. Though they shared a common linguistic group, New England's Indigenous peoples belonged to dozens of autonomous communities, each with diverse forms of cultural expression, social and political organization, and subsistence production. All of New England's Algonquian-speaking communities recognized, however, that community stability and tribal well-being required sophisticated adaptations to local environments. Indigenous peoples inhabited, and drew their subsistence from, every ecological niche. Anthropologist Kathleen Bragdon suggests a precolonial Native geography, what she terms a "tripartite settlement" model, in which different Indigenous communities occupied coastal, riverine, or upland village sites.[5] To expand and diversify the resource base, many Algonquian communities moved frequently between these three environments on a seasonal basis. Wabanaki family bands traveled downriver in spring to fish for alewives, shad, salmon, and sturgeon before returning to inland planting grounds to sow corn, squash, and beans. By the midsummer, they removed to the coastline to pursue seals and porpoise and to gather shellfish. After harvesting their crops, clans moved once again to hunting and trapping grounds deeper in the interior river valleys.[6] This seasonal movement required intimate knowledge of wildlife migrations, plant life, and soils so as to capitalize on the fixed resources available within each particular ecosystem. Alternatively, Bragdon notes that abundant coastal and estuarine resources helped encourage what she terms a "conditional sedentism." By the late Woodland Period (AD 1000), Indigenous communities also pursued a "riverine village-based sedentism," where soils could support a productive polyculture of maize, beans, and squash.[7] These interpretations line up well with early colonial descriptions of Algonquian subsistence practices. European colonists documented tribal movement between semipermanent village sites and outlying settlements associated with fishing, hunting, gathering, and horticulture. These smaller village sites may have been occupied part of the year, or sometimes year-round, which indicates a division of labor within Indigenous communities that helped support the village, clan, or tribe as a whole. Examples of this boundary settlement activity might have included berry gathering, maple sugaring, timber harvesting, canoe making, shellfish processing, and fishing.[8]

The English colonist Thomas Morton recorded some of the earliest ethnographic observations related to New England's Indigenous inhabitants. Released in three volumes beginning in 1637, *The New English Canaan* revealed not only Morton's deep animus for his dogmatic Puritan neighbors but also his genuine interest in understanding Algonquian culture on their own terms. Soon after Morton landed ashore in the recently founded colony of Plymouth in 1622, he encountered "two sortes of peoples, the one Christians, the other Infidels." He declared the Indigenous people the "most full of humanity, and more friendly than the other." Morton established a settlement and fur trading post, named Ma-re Mount, and followed a far more diplomatic path toward a mutual accommodation with the local Indian peoples. Taking note of their traditions and customs, Morton documented the seasonal removal to satellite encampments: "They use not to winter and summer in one place, of that would be a reason to make fuel scarce; but . . . remove for their pleasures; sometimes to their hunting place, where they remain keeping good hospitality for that season; and sometimes to their fishing places, where they abide for that season likewise."[9] Morton rightly maintained that this practice served to spread the environmental impact over a large area and therefore limit the exhaustion of any single ecosystem. An extensive resource base was a crucial element of survival, but these practices were also informed by Algonquian cosmologies that promoted sustainable relationships with the environment.[10]

Sustainable resource use required intimate knowledge of the environment, and according to early British colonists, New England's Algonquian communities possessed a deep understanding of everything piscatorial. William Wood spent four years with the Puritans around the founding of the Massachusetts Bay Colony, returning to England sometime in 1633. During this relatively short colonial venture, he took note of "expert" Indigenous fishermen, "experienced in the knowledge of all baits" and "knowing when to fish in rivers, and when at rocks, when in bays, and when at seas." Before the arrival of British colonists, Algonquian fishermen used line crafted from "hemp more curiously wrought, of stronger materials than ours, hooked with bone hooks."[11] According to Wood, they understood the timing of seasonal migrations, the appropriate techniques to ensure a healthy catch, and the materials necessary to implement those techniques. He and other Europeans held this knowledge, accrued over thousands of years, in the utmost esteem. The tenuous nature of first-generation colonial settlement led to a recognition that survival and stability would be in part predicated on borrowing from Algonquian subsistence strategies.

In addition to these colonial accounts, the remains of prehistoric fishing camps attest to the vitality of an Indigenous fishing culture. Typically associated with the remains of an ancient weir, seasonal fishing camps have been found scattered throughout the Northeast. On the Merrimack River alone, three camps—the Weirs Beach, Neville, and Buswell sites—are spread throughout the entire watershed, from headwaters in northern New Hampshire down toward the coastline. At the LeBeau site, beside the Quinebaug River in southern Connecticut and within the ancestral territory of the Pequot and Mohegan Nations, a crew of researchers recovered over eight thousand artifacts, including stone tools, projectile points, clay pottery, net-sinking stones, and fire-cracked rocks. The sheer quantity of items alone indicates that this seasonally occupied fishing camp was both extremely productive and a fixture of tribal subsistence. Researchers suggest that Indigenous fishermen used the fire-cracked rocks to cook or preserve the catch to support a larger network of Indigenous villages. The relatively small cutting tools found at the site also indicate that women played a prominent role in processing and transporting the catch to the central village.

Lands within the riparian floodplain were also favorable for maize horticulture and could further encourage a conditional sedentism that minimized frequent removal to additional resources.[12] Fall lines along the region's large watersheds were particularly popular gathering sites and offered some of the most productive fishing and farming grounds, as significant changes in elevation could substantially slow and restrict the passage of anadromous fish upriver. Native American campsites were situated in a strategic manner to maximize exploitation of an extensive resource base. Colonial records support this claim. British colonists purchased lands from Native peoples that included villages along the Merrimack River at present-day Newbury, Amesbury, Haverhill, Lowell, and Chelmsford. These Indigenous riparian settlements extended up the Merrimack Valley as far north as Lake Winnipesaukee.[13] Previously home to countless generations of Native Americans, these riparian lands were almost certainly acquired as a result of the demographic catastrophe that followed waves of epidemic disease and the aggressive territorial expansion that characterized settler colonialism.

The remains of several ancient weir sites across the region speak to the central position of river fishing within Indigenous material and cultural life. Along with colonial accounts of abundance, it seems that river fisheries sustained Indigenous communities without fundamentally undermining the resource by the time of European contact, the product of an environmental worldview that privileged the long-term productivity of lands and waters that supported the health and well-being of the tribal community. Whether this was the product of intentional design or the by-product of decentralized sociopolitical systems and low population densities has been a matter of some controversy in academic circles. Perhaps Indigenous populations were not so large as to pressure or stress the resource base and significantly diminish fish stocks and terrestrial wildlife.[14] The absence of a commodity-based economy would have encouraged Indigenous communities to live within the carrying capacity of the land. But scholars have increasingly emphasized the values of stewardship, reciprocity, and culturally defined appropriate conduct at the foundation of Indigenous environmental knowledge.[15] Dispensing with the trope that all American Indians lived in harmony with nature at all times, Michael E. Harkin and David Rich Lewis write in *Native Americans and the Environment: Perspectives on the Ecological Indian* that historically, "most Native American cultures demonstrated gross-level sustainability—the ability to persist in the same environment over millennia, although not necessarily at the same population level."[16] To persist in the same environment over such long periods of time, one must be attentive to the resources and ecological rhythms that allowed for a comfortable subsistence. Traditional ecological knowledge systems could have helped discourage harvesting more fish than was considered necessary.[17] It is most likely that a combination of factors fostered an environmental sensitivity that allowed Indigenous peoples to fish, hunt, farm, gather, and inhabit the Northeast for thousands of years before European colonization. Whatever level of sustainability the region's Algonquian peoples managed to achieve prior to European contact, colonization would soon unleash radical ecological transformations.

River Traditions: Continuity and Competition

When the British first crossed the Atlantic and arrived at the shores of New England, they encountered a largely foreign environment, but important ecological connections helped solidify their place in what they understood, from their narrow cultural perspective, to be a "new world." As historian Jeffrey Bolster wrote, "The living ocean along the northeast coast of America mirrored Englishmen's coastal ecosystem at home."[18] Aquatic environments brought some welcome familiarity. Salmon in particular provided some necessary ecological continuity for European colonists. Atlantic salmon fishing had existed for centuries in Great Britain, with Scotland particularly well known for both an abundant salmon fishery and local commitments to the survival of the resource. Scottish officials regulated the salmon fishery as early 1406.[19] In addition, English common law had long recognized the importance of anadromous fish runs by forbidding any subject from obstructing their passage upriver.[20] In this sense, river fishing occupied a special place within both Indigenous and European cultures. This shared tradition would simultaneously represent a point of common interaction and tension as colonial settlements spread through the region and brought about deep shifts in local ecology.

FIGURE 2. Engraving from *Ballou's Pictorial Drawing-Room Companion*, 1855. Native American and colonist both fishing at Weirs Channel, NH. A waterfall can be seen behind the Native fisherman, whereas on the other side, a small mill has obstructed the natural course of the river.

Accounts from early explorers, settlers, and colonial leaders invariably praise New England as a veritable Garden of Eden, with resources sufficient to sustain, and indeed encourage, colonial migration. John Smith famously described a section of coastline near present-day central Maine: "From Penobscot to Sagadahock this coast is all mountainous and isles of huge rocks, but overgrown with all sorts of excellent good woods for building houses, boats, barks, or ships, with an incredible abundance of most sorts of fish, much fowl, and sundry sorts of good fruits for man's use." Smith provided ample reason for settlers to consider New England, perhaps painting too rosy a picture given the harsh reality of carving working settlements from the wilderness, but there was no doubt that resources outside the agrarian land base, such as fish and timber, could make this prospect a little less daunting. According to Smith, the "carpenter, mason, gardener, tailor, smith, sailor, forger, or what other . . . though they fish an hour in a day, to take more than they eat in a week; or, if they will not eat it . . . sell it, or change it with the fishermen or merchants for anything they want."[21] Here, Smith anticipates the position of fish within an internal colonial economy of barter and exchange, a tradition that provided opportunities to both satisfy household needs and engage local markets. With modest effort, one might secure a week's supply of food for consumption or trade. He also describes coastal fishing in the language of the commons, a resource available and open to all colonists, and hints at the development of a localized economy geared to household independence. Whether a carpenter, mason, farmer, or the like, all could improve their economic prospects through part-time, or seasonal, fishing. This passage evokes a feeling of shared prosperity partially supported by coastal and freshwater ecosystems. River fishing, much as Smith anticipated, would continue to support New England's rural economy long after the colonial period.

Subsistence fishing represented an essential component of the economic landscape, and Smith rightly understood that this account would attract considerable attention from those considering the perilous transatlantic journey. In the end, his evocation for emigration was clear: "So freely hath God and his majesty bestowed those blessings on them that will attempt to obtain them, as here every man may be master and own labor and land, or the greatest part in a small time."[22] The widespread availability of land, combined with the resources necessary to support a comfortable subsistence, represented an attractive alternative to poverty, resource scarcity, and overpopulation in England.

During the first several generations of British colonization, settlers began to push the boundaries of their coastal settlements, often following river systems into the forested interior. According to Thomas Hutchinson, a colonial governor and historian, colonists encountered groups of "Northern Indians," inhabiting "many distinct settlements on the lesser channels of the Piscataqua or Newichewannock [Salmon Falls] River." Hutchinson also describes a "noted plantation" at Newbury Falls on the Parker River that featured "plenty of fish there at all seasons."[23] On the Merrimack, situated between these settlements on the Piscataqua and the Parker Rivers, Indigenous communities inhabited villages on both sides of the estuary. Bands of "Wainooset, Patucket, Amoskeag, Penicook" occupied sites "from the mouth fifty miles or more" upriver. The eighteenth-century historian Jeremy Belknap notes that this expansion into Indigenous territory continued when, in 1726, colonists from the lower Merrimack Valley journeyed north to the Province of New Hampshire to settle along the river at Penacook Plantation, a tract of land that would later include Concord. Belknap writes that these lands had once "been the seat of a numerous and powerful tribe of Indians." As Native American populations declined with the spread of disease, warfare, and dispossession, the Provinces of New Hampshire and Massachusetts Bay chartered new townships between the Connecticut and the Merrimack Rivers.[24] Competition for land and resources aggravated already tense relations between remaining populations of American Indians and British colonists. Though the early stages of settler colonization were, according to Bruce J. Bourque, "generally peaceful, if not always friendly," this era of good feelings faded "as Europeans began to clash over territorial claims and access to furs, often using liquor and firearms to enlist Native support, and resorting to other abusive tactics."[25] Sadly, Native Americans found themselves caught between European nations competing to dominate a colonial geography once defined by Indigenous power. British colonists engaged in fraudulent and patently unfair land transactions, usually undertaken without tribal consensus.[26] In 1635, the General Court of Massachusetts banished Roger Williams in part because he dared to profess the radical proposition that American Indians possessed their lands "by right of first occupancy."[27] Conflict was not exclusive to questions of colonial dispossession and land tenure; colonial encroachment on traditional Indigenous rights to ancestral waters signaled a continued deterioration in relations.

As European settlements spread across the Northeast, nearly every village constructed a gristmill and a sawmill powered by water from a small dam. Seventeenth-century mill dams were modest in size but still capable of disrupting migratory fish runs.[28] In addition, as mill dams blocked access to spawning habitat, lands cleared for tillage compromised Indigenous hunting grounds, and domesticated live-stock destroyed essential habitat for native wildlife.[29] Belknap wrote that as Indigenous peoples began to understand the consequences of their relative tolerance of European neighbors, "they repented of their hospitality, and were inclined to dispossess their new neighbors, as the only way of restoring the country to its pristine state, and of recovering their usual mode of subsistence." But there would be no turning back the clock, as the first fifty years of European settlement brought about a demographic catastrophe for Algonquian peoples. Disease ravaged Indigenous communities with particularly devas-tating epidemics in Massachusetts in 1616 and then west toward the Connecticut River Valley in 1633. In some cases, mortality rates reached some 90 percent.[30] Population collapse limited the prospect of any meaningful pushback against European monopolization of essential resources such as fish.

American Indians fought to retain lands unjustly transferred to colonial hands, as well as a spiritual regime that restrained unmit-igated environmental exploitation. At the same time, colonists worked to integrate Indigenous hunters, fishermen, and trappers into networks of regional and transatlantic commerce. To acquire European goods, many Indians participated in the fur trade and contributed to the near extinction of fur-bearing mammals across the Northeast. While many Indigenous communities endeavored to maintain traditional relationships with the environment, as epi-demic disease stormed through the countryside, undermining cen-turies of environmentally sensitive cosmology, colonization fractured relationships with the land once defined by animistic worldviews and reciprocity.[31] The knowledge and techniques that informed centuries of an Indigenous fishing culture—the location of productive fishing grounds, the variable timing of seasonal fish runs, the material and equipment necessary to produce a successful result, and the valued riparian settlements where a combination of horticulture and fishing supported a network of village communities—transferred to eager colonists, who adapted river fishing as part of their own subsistence and commercial designs.

Confronting the Myth of Abundance

Not long after the founding of Massachusetts Bay in 1628, European assumptions of unlimited abundance ran up against the inescapable reality of environmental decline. The cornucopian vision of virtually inexhaustible resources belied the ecological transformations that would soon accompany settler colonialism. Motivated by the pursuit of land and wealth, Europeans introduced new sets of environmental impacts associated with private ownership, commodity extraction, and transatlantic commercial networks. A nascent capitalist ethic interpreted the natural world through the lens of commerce, which often resulted in environmental disaster. Beavers, highly desired in Europe for their fur, fell victim to this ruthless acquisitive spirit. Colonists were amazed by the ubiquitous presence of these furry creatures, but by the late eighteenth century they had all but disappeared.[32] Similar environmental disruptions—deforestation, river siltation, soil compaction—took far less time to manifest. The decline of fish resources was a function of similar market pressures.

John Cabot's 1497 exploration of the Northwest Atlantic and the Grand Banks of Newfoundland unleashed centuries of English, French, Spanish, Basque, and Portuguese competition to capture the rewards of this exceptional marine ecosystem. By the late eighteenth century, the once-abundant Northwest Atlantic cod grounds began to show signs of overfishing.[33] Localized depletions within coastal and freshwater habitats appeared much earlier. Colonists quickly built weirs on tidal flats up and down the Massachusetts coast. William Wood wrote in *New England's Prospect* that a weir on the Charles River captured hundreds of thousands of shad and alewives over the course two tidal cycles.[34] This kind of intensive fishing provoked a series of interventions by governing authorities. Between 1639 and 1650, colonial magistrates required fishermen to open their weirs two days a week, prohibited farmers from fertilizing their fields with striped bass or cod, and revoked a lease to fish for striped bass on Cape Cod Bay. Despite these initial reforms, colonists continued to identify disturbing trends. In 1673, townsfolk near the Merrimack River were so dismayed by the decline of sturgeon that they petitioned authorities to close the fishery to the general public and instead issue a limited number of special licenses. In 1664, colonists in Taunton denounced a sawmill for obstructing the entire seasonal run of alewives.[35] Coastal fisheries showed similar signs of stress, which led Massachusetts Bay to take additional steps to restrict mackerel fishing in 1670 and again in 1682.[36] In these examples, fast-growing colonial populations exploited their local environments to the breaking point, with leadership stepping in to moderate the fishing activity. Demographic pressures would test the strength and efficacy of the magistrate's preliminary experimentation with resource conservation.

Steady population growth throughout the seventeenth century helped accelerate the environment transformations unleashed by European colonization. Philip J. Greven's demographic study of colonial Andover, Massachusetts, a town that borders the Merrimack River, notes that "the last two decades of the seventeenth century and the first decade of the eighteenth century constituted a period of unparalleled growth in the population of Andover."[37] As early as 1705, the community numbered one thousand inhabitants. This placed enormous pressure on fathers who sought to pass on independent livelihoods predicated on access to land. After a little more than 150 years of settlement, Andover confronted a scarcity of arable land.[38] There was too little to divide among the male progeny to ensure the next generation could meet their needs. This unwelcome development underscores the environmental pressures attached to such rapid population growth that had precipitated an increasingly intensive use of the region's land and waters.

Just as Andover grew to over one thousand inhabitants, residents from the wider Merrimack valley expressed concerns about the health of the river's fisheries. Persistent complaints related to weirs, milldams, and other obstructions led the Province of Massachusetts

Bay to institute a more comprehensive system of oversight to prevent overfishing. In 1709, the colony ordered that only persons who had sought an allowance from the "general sessions of the peace," in each respective county, could legally use a weir or similar trapping device. In essence, the law transformed the fishery into a closed system that empowered local officials to issue licenses on the basis of whether such an allowance constituted a "public good or damage" to the same and set a compelling precedent that would be continually referenced by fishermen and sympathetic legislators for decades to come.[39] Open access to a staple of the New England diet fell firmly within the public good, but only if fishermen harvested responsibly. The communal principles that ensured the right to access the seasonal harvest of river fish was also widely understood to be worthy of protection, but not at the expense of the commons itself. As such, the law declared any weir or obstruction installed in a river without the approval of colonial authorities a "common nuisance" that would be "demolished and pulled down, not to be again repaired or amended." In 1727, lawmakers attached a ten-pound penalty for weir fishing without the "approbation of the courts of the general sessions of the peace in the respective counties," designed "for the more effectual preventing such nusances [sic]."[40] With these two measures, the Province of Massachusetts Bay embarked on the first legislative program to curtail overfishing, but weirs did not represent the sole impediment to sustainable seasonal fish runs.

New England's first water-powered sawmill opened in 1634, and over the next century rural mills spread across the countryside, lining the banks of streams and brooks that otherwise led to upriver spawning sites for anadromous species.[41] In the 1740s, lawmakers first attempted to confront the potential negative impacts for migratory fish runs by requiring "a sufficient passageway" where any dam crossed "such rivers or stream where salmon, shad, and alewives, or other fish usually pass up into the natural ponds to cast their spawn."[42] But this seemingly reasonable decision caused many mill operators to question what exactly constituted a "sufficient passageway." In addition, the timing of upstream migrations varied across

species and location within the watershed, making it difficult to identify the appropriate moments to open these passages. Since milling
and fishing were both indispensable to the growing colonial economy, legislators set out to find a middle way. Colonial authorities
empowered the "justices of the court" to select three "disinterested
persons" who would determine the appropriate dimensions, tailored to each dam, that would accommodate the free passage of fish
upstream. Those disinterested individuals were also charged with
determining "how long the same shall be kept open."[43] Dissatisfied
with this compromise, millers continued to protest by denying the
presence of migratory fish within their streams.[44] As with overfishing, the Massachusetts General Court constructed legislative remedies to balance competing interests to these waters, but clearing the
way for seasonal fish migrations would represent an ongoing project. A compromise that would satisfy all interested parties remained
elusive, and legislators continued to position themselves as neutral
arbiters of the public good. By the late eighteenth century, effective
compromise would need to account for the specific ecological factors
at play within each freshwater ecosystem.

Between 1709 and 1745, overfishing and milldam obstructions
provoked a series of colonial measures to protect river and coastal
fisheries. In 1741, the Massachusetts General Court began to char-
acterize this work as the "preservation" of "alewives and other fish."
They believed this could be accomplished only if millers and fishers allowed some fish "free passage up and down the rivers in their
seasons."[45] This rhetoric of preservation set an essential legislative
foundation that would endure for the next hundred years and come
to represent the guiding light for an emerging freshwater conservation agenda. By 1781, both petitioners and legislators consistently
returned to the mantra of preservation when drafting and revising
inland fisheries regulations. Indeed, for many this preservation mission represented a call to action; meaningful collective effort would
be required to prevent the utter collapse of inland fisheries.

Native Americans fished the inland waters of the eastern woodlands for thousands of years, seemingly without compromising the

resources that supported tribal subsistence. The precontact and early colonial periods provide a glimpse into the significance of rivers and fish resources amid cultural and ecological transformations. Archaeological discoveries indicate that rivers, and the fish found therein, contributed substantially to Indigenous subsistence in the precontact era. European settlers were astounded by North American fish stocks and equally impressed with the skill with which Native Americans fished coastal and inland waters, quickly adopting similar fishing practices. What they failed to adopt was a worldview that served to restrain the aggressive fishing practices that began to threaten this resource only decades into the colonial experiment. Early in the colonial period, new economic forces disturbed the ecological balances that had once sustained the region's river fisheries. This marked a turning point where fishermen realized they would need the help of governing authorities to protect future returns. These farsighted colonial petitioners would come to represent the backbone of grassroots efforts that sustained the fisheries across multiple generations and ensured that river fish would continue to support rural agricultural communities well into the nineteenth century.

CHAPTER THREE

FARMER AND FISHERMAN

Seventeen miles up from Casco Bay on the coast of Maine, six rivers converge to form the largest freshwater estuary north of Chesapeake Bay. Merrymeeting Bay, a cheerful English place-name for this inland delta, empties through a narrow, hazardous corridor known as "the Chops" into the lower Kennebec River before finally reaching the Gulf of Maine. For countless generations, the Indigenous Wabanaki inhabited the Kennebec watershed, but beginning with the ill-fated Popham Colony in 1607, both British and French colonial interests worked to secure a foothold within Maine's rugged coast. Harsh winters, challenging soils, a century of imperial conflict, and Wabanaki resistance stunted British ambitions in the region until the end of the French and Indian War in 1763. Massachusetts settlers migrated to the town of Bowdoinham, incorporated in 1762, to farm the comparatively rich freshwater marshes on the western shores of the bay, above the mouth of the Cathance River. On the eastern boundary of the bay, the town of Woolwich forms a large peninsula where settlers erected grist- and fulling mills but also fished for alewives in the lower Kennebec, the Sasanoa, and the Back Rivers. The Androscoggin and the Kennebec Rivers meet further south, carving the landscape into a series of jagged and fragmented islands and peninsulas. With slow population growth well into the early nineteenth century, townsfolk around the bay enjoyed countless locations to fish these waters free from the demographic and commercial pressures that depleted densely settled riparian environments across southern New England. Elias Robbins was one of several farmers fortunate enough to inhabit this exceptional freshwater environment.

In May 1816, Robbins waded into the "great flats" of Merrymeeting Bay somewhere near Bowdoinham to construct a brush weir. Like any seasoned fisherman, Robbins failed to disclose the precise location of his choice fishing spot, but over the next several months he carefully documented trades, sales, and debts satisfied with alewives and salmon.[1] A yeoman farmer from Woolwich and later Bowdoinham, Maine, Robbins would continue to meticulously record the day-to-day transactions that defined his economic life over the next thirty years. A jack-of-all-trades, by the end of his life Robbins had amassed a respectable farm where he grew corn and oats, raised sheep, and harvested timber, but each spring and summer Robbins turned his attention to the tidewater flats of Merrymeeting Bay and the seasonal runs of anadromous fish. Altogether, his account book underscores the patterns of seasonal labor at the heart of New England's agrarian economy, the endless pursuit of household autonomy, the formation and maintenance of cooperative trade relationships with the broader community, and the central role of river fishing for rural economies of semisubsistence, barter, and exchange. The inscription on the first page of the book, likely by a relative who later came to possess the document, reads: "Elias Robbins, Woolwich, Maine. Farmer and Fisherman."

Blessed with abundant freshwater, extensive river systems, and prolific inland and coastal fisheries, seasonal fishing fit seamlessly into the rhythms of agricultural labor and the pursuit of economic security across the New England countryside. Shortly after farmers finished planting their crops in the spring, anadromous fish began to leave the sea toward freshwater spawning grounds in such numbers as to astonish the casual observer, and although New England's rural economy was overwhelmingly rooted in the land and soil, hunting, fishing, gathering, and household manufactures offered additional pathways toward a comfortable subsistence. For farmer-fishermen like Elias Robbins, the predictable influx of migratory river fish represented an important component of a diverse and multi-occupational economic landscape geared primarily toward subsistence production and local exchange. James A. Henretta notes that for preindustrial farmers, the "dimensions of economic existence" might vary from one region to the next depending on the accessibility of domestic or transatlantic commercial networks, but cultural and environmental constraints equally limited the reach of the marketplace.[2] Faced with unpredictable weather patterns, marginal soils, a relatively short growing season, and the prohibitive cost of overland transport, most households chose the security of an extensive farm ecology through which forests, fields, pastures, meadows, orchards, and waters underwrote a modest prosperity and an egalitarian social climate.[3] Account books, diaries, and nineteenth-century oral histories describe farmer-fishermen busy providing for the immediate needs of their respective households and operating within an economic culture that emphasized production for consumption and local trade. They fished to supplement the household diet, preserved the rest for leaner times, exchanged their catch for goods and services within the community, and relied on river fishing to overcome moments of scarcity. When commercial pressures coalesced to threaten the vitality of this resource, this rural economic context set the stage for a defense of traditional fishing and an economic culture predicated on a diverse and diffuse resource base. Because this resource occupied such a prominent position within the agrarian economy, farmer-fishermen embraced conservation initiatives designed to pass along a sustainable environmental heritage to the next generation.

At the turn of the nineteenth century, a great majority of New Englanders inhabited rural communities of small independent farms—living representations of Thomas Jefferson's vision of an agrarian republic. Historian Diana Muir writes that "in a New England village, every miller was at least a part-time farmer—as was virtually every minister, lawyer, blacksmith, and storekeeper—and the farms they worked produced a very small surplus, when they produced a surplus at all."[4] Striving for stability, security, and independence, yeomen farm households embraced a system of mixed husbandry, a land use strategy that combined livestock, woodlots, pasture, and croplands to the maintain the fertility of planting fields.[5] Elias Robbins began his account book in 1815, but by the time his name was listed in the federal census, he appeared to have followed a similar path toward a comfortable independence, what early Americans themselves termed a "competency."[6] In 1850, fifty-seven-year-old Robbins worked his Bowdoinham farm with his wife and seven children, though two of his older sons worked as carpenters in the local shipbuilding trade.[7] The Agriculture Census reports that Robbins's farm constituted one hundred acres of "improved" land, another thirty acres "unimproved," three milk cows, two working oxen, and twenty sheep.[8] Robbins's improved land would have included tillage and pasture for his animals, while he most likely kept "unimproved" tracts as woodlots. With this extensive farm ecology that featured lands in various stages of improvement, Robbins harvested thirty bushels of corn and forty bushels of oats. The census estimated the total worth of Robbin's farm at two thousand dollars, well within the average for the larger Bowdoinham farming community.[9] Across the Northeast, local economies comprised of relatively equal landowners, and an egalitarian social climate would shape attitudes toward access and distribution of critical natural resources.

From 1815 to 1850, Robbins pursued this generalist strategy to great effect. Slow population growth along the Kennebec River valley afforded Robbins and his neighbors the extensive land and resource base to forestall the rural land crises that plagued long-settled towns further south. In 1800, only six hundred individuals lived within Bowdoinham proper.[10] As Brian Donahue has shown in his study of colonial Concord, Massachusetts, demographic pressures in the mid-eighteenth century undercut an intensive but sustainable system of mixed-husbandry adapted to the ecological limits of the region's soils and climate. With towns that had more space to accommodate agricultural expansion, Carolyn Merchant describes how colonists carefully hewed to an extensive farm ecology that combined polycultures, three-field crop rotation, upland pastures for livestock, and long fallows that cycled between woodlands and planting fields.[11] While these subsistence agricultural regimes were not without their own internal contradictions, both Merchant and Donahue identify market capitalism as the key turning point for environmental degradation in the Northeast. Growing populations, combined with specialization, pressed hard on the land and led to declines in farm size, deforestation, and, eventually, soil exhaustion. Yet, in a specific time and place, communities of yeomen farmers like Robbins enjoyed a comfortable independence predicated on family labor, access to arable land, and a broad-based, seasonally variable collection of natural resources.[12]

Whether an extensive or intensive farm ecology, the quest for economic independence was not a solitary effort. Yeomen farmers also benefited from a cooperative economic culture and vibrant networks of local exchange that further supported the quest for household independence and economic security. In between cultivating his land, tending to his sheep, and fishing the great flats of Merrymeeting Bay, Elias Robbins lent his services to neighbors and friends in exchange for much-needed goods that he could not produce within his own household. He brought his oxen to neighboring farms to plow each spring in exchange for additional hay to feed his sheep. Something of a handyman, Robbins fixed broken stoves, windows, and torn bass nets. He butchered hogs, cut wood, set bricks, and harvested potatoes.[13] Providing a seemingly endless variety of services, Robbins must have been an indispensable member of the Merrymeeting Bay community. It seems that for nearly any occasion, Robbins was there to lend his expertise. Far from striving for total economic independence, most farmers participated in these exchanges of goods and labor that supported a comfortable independence.[14] True self-sufficiency would have represented a difficult prospect; informal divisions of labor within these agrarian communities furnished necessary skills to households in need. Carpenters, shoemakers, millers, tanners, and the like all provided services to friends, neighbors, and other members of the extended social network.[15] This system of rural barter and exchange aligned with the republican ideal of a free and autonomous citizenry, as families bargained with one another with the knowledge that they would retain control over their own labor and, ultimately, their economic fortunes. In this framework, semisubsistent households frequently exchanged goods and labor, highlighting the importance of extended kin networks and community-based webs of mutual cooperation.[16] New England's yeomen farmers, a mostly literate population, kept detailed account books, journals, and diaries that documented decades of community exchange in goods and labor.

From 1815 to 1845, during the spring and summer runs of migratory species, Robbins fished and then bartered his catch with members of his Merrymeeting Bay community. An exceptionally capable fisherman, he recorded exchanges with bass, salmon, alewives, and shad. His account book reveals significant local demand for his catch. In May 1816, Robbins settled an outstanding account with Mattias Fisher totaling $2.40 with seven pounds of salmon and nine shad.[17] He traded alewives in much larger quantities, since this species would have been especially abundant within this rich freshwater estuary. In July 1816, he sold 100 alewives to Francis White and the same amount to Phillip White the next month. In October of the same year, perhaps seeking to unload the remainder of the season's catch, he parted with 280 alewives for $2.80.[18] It is unclear whether Robbins accepted cash payments or if these men later settled their debts with barter, but the account book also provides evidence for more direct, reciprocal transactions of goods and services. In May 1826, in the early stages of the spring fishing season, Robbins acquired two bushels of corn from Elihu Hatch for nineteen and a half pounds of salmon.[19]

Accounts between Robbins and his associates could sometimes drag on for several years, which would have been of little concern given that New England's farmers inhabited an intimate face-to-face society based on personal reputation and good faith. In 1827, Robbins agreed to provide shad, hay, and the use of his oxen for a half day, among other items, to Zehlin Prebble in exchange for two days of labor. Prebble finally settled his outstanding balance five years later with a bushel of corn delivered in 1832. Outside these exchanges of agricultural goods and services, Robbins also used his catch in transactions with local tradesfolk. In 1829, Robbins settled a debt with his neighborhood blacksmith with thirteen shad, a half gallon of vinegar, and fifty bushels of coal.[20] Lastly, Robbins lent his fishing expertise to other community members also engaged with the local fishery. In April 1836, he spent five days hauling brush and laying out a weir for Harry Prible. For his trouble, Robbins received a share of the catch, some 825 alewives.[21] Along the coast and inland river valleys, farmer-fishermen entered the river commons not only to provide for their household needs but also to acquire goods and services that they were unable to produce or supply within the household. At the foundation of this system was an intimate sense of reciprocity, obligation, and mutual cooperation, with fish serving as a key mechanism that solidified and strengthened social connections between community members. Even those who never put a line in the water, dragged a seine, or set a weir could rely on such fishers as Elias Robbins to distribute the catch throughout Merrymeeting Bay. These networks of community exchange would continue to fuel a dynamic and productive rural economy for decades to come. Reuben Devereaux, on the eastern side of Penobscot Bay near the town of Castine, recorded community exchanges of alewives, salmon, pollock, and cod into the 1870s.[22]

With the arrival of migratory fish, nearly every household could expect a fair share, provided either as part of a cooperative arrangement or by a lone fisherman utilizing waters located on his own property or within the boundary of a respective town. On the Merrimack River, close to the celebrated fishery of Amoskeag Falls, Matthew Patten partnered with friends and family to acquire his share of this seasonal resource. Born in Northern Ireland in 1719, Patten migrated with his family to New England in 1728. The Pattens soon put down roots in Bedford, New Hampshire, where Matthew and his brother secured a small section of land on the periphery of British colonial life. Though the Patten clan represented but one of many Scotch-Irish communities to seek a better fortune in British

North America, his story is far from ordinary. He rose through the social ranks to serve as a judge of probate, represented the towns of Bedford and Merrimack in the General Court of New Hampshire, served briefly as a member of the governor's council, and was finally appointed justice of the peace in 1751, an office he occupied until his death.[23] By 1774, Patten found himself swept up in the cause of the American Revolution and joined his local committee of safety to harass and punish loyalist officials who interfered with the election of delegates to the Continental Congress. Throughout the course of the war, Patten participated in numerous "insurgent" activities designed to advance the patriot movements in his southern New Hampshire community.[24] While Patten negotiated conflicting identities as a British colonial public servant and then energetic revolutionary, he pursued multidirectional pathways to secure a competency. From 1754 to 1788, Patten maintained a sizable farm, harvested timber, trapped fur-bearing mammals, worked as a carpenter and surveyor, and fished the Merrimack River every spring and summer.[25]

Patten's record of the seasonal fishery at Amoskeag Falls provides a window into one of the most historically productive river fisheries in the region and the social character of spring fishing as massive runs of shad struggled to ascend the cataract in search of spawning grounds further north near Lake Winnipesaukee. Patten's success at Amoskeag Falls often depended on his participation in informal "fishing companies," essentially cooperative agreements among friends and neighbors to pool material and labor in exchange for an equal share of the catch. On June 2, 1762, Patten ventured off to the falls and spent the next four days fishing for shad and salmon. On the third day of his trip, he decided to join with "the company the other Robert Macmurphy had & I pd Mr Russ my fishing expenses." The next day, Patten paid another member of the company his share of the "fishing Expenses," before making the journey home with "49 shad and 2 halves of salmon the rest of the shad I sold and I sold 22 shad to James Aiken of Chester . . . and he would not pay me." Perhaps still irritated with this ill treatment, Patten nevertheless went back to Amoskeag Falls on June 11 and, presumably fishing by himself, "catched a salmon of 11 ½ £." Later that month, Patten once again joined his company at a new fishing spot set above the falls they called "the pulpit" and hatched a plan to improve their access to the water. On June 17, Patten and his group "went to work at Amoskeag falls half the day . . . to blow a place to fish at above the common setting, we bored two holes a little into the rock in order to blow and beat off a considerable quantity with a stone hammer."[26] In search of more productive fishing grounds, Patten and others altered the environment, as they had in their agricultural pursuits, to increase their yields.

During this one season, Patten alternated between collaborative and solitary fishing strategies that involved specialized gear and methods tailored to available labor inputs and the quarry under pursuit. Historian Daniel Vickers notes that when Patten fished by himself for shad, he most likely used a single scoop net—a small net situated at the end of a long wooden pole. Amoskeag shad fishers positioned themselves on the banks of the Merrimack, submerged the net near the falls, and "scooped" fish from the water onto the ground.[27] While not the most efficient fishing method, the scoop net required the least amount of preparation, resources, and effort to enter the shad fishery. Conscious of the fleeting opportunities to capitalize on this seasonal resource, fishermen such as Patten chose to pool labor and resources to substantially increase their returns.

Patten described instances in which his company used either scoop nets or seines but failed to lay out the specific cooperative strategy associated with the respective equipment. Here, Sylvester Judd's exhaustive nineteenth-century town history of Hadley, Massachusetts, helps fill in some of the gaps. In 1848, Judd interviewed two seasoned fishermen with decades-long experience of the annual shad run at South Hadley Falls on the Connecticut River in western Massachusetts. Joseph Ely and Justin Alvord recalled that "boats were drawn to places on the rocky falls, fastened, and filled with shad by scoop-nets; then taken ashore, emptied and returned." The two fishermen estimated that a skeleton crew of two men could take "2000 to 3000 shad in a day."[28] Further below the falls, companies fished with seines: nets up to three hundred feet long and twenty feet deep. Vickers describes the process in which Patten and his compatriots would attach "one end to a pole or rock on the bank, then coiling the remainder inside a small boat and paying it out over the stern or 'shooting' it while they rowed the boat in a semicircle into the river and back to shore." The men then weighted the net to the river bottom, buoyed the top to the surface of the water, and hauled both ends back toward the shore, a process that, Vickers notes, might have lasted several hours. Patten's diary clearly shows that either cooperative strategy promised a larger return. On June 11, 1764, Patten took 361 shad from the river. During Patten's scattered visits to Amoskeag Falls in early June 1765, he took home 433 shad.[29] When the salmon began to enter freshwater systems a few weeks after the shad—in schools that were smaller and less concentrated than their anadromous counterparts—Patten and his company might drag a seine below the falls or, alternatively, install a fixed setnet designed to enmesh fish as they swam upriver. With his catch on shore, Patten would then preserve his share for later use, barter some for goods, or sell some to farmers that also traveled to the falls for shad season. On June 10, 1772, Patten notes that his sons "sold 61 from their setting place," while he "sold on the Island 184 for money and 20 I let Rob McGregor have for a bushel of salt."[30]

Similar to Elias Robbins's rural community further north in Maine, migratory fish, especially shad and salmon, represented an important article of trade in the Bedford economy. For the most part, Patten used fish to acquire goods and services and to settle debts, with the rest preserved for future use. During a four-day stretch in May 1766, Patten took home 550 shad and then traded 120 fish for two bushels of salt. Patten's duties as justice of the peace involved regular travel to record testimony and take depositions. During one such trip he lodged at the home of a Mrs. Osgood and payed for his stay with a large salmon.[31] Robbins's and Patten's dealings in fish almost always took place within this local context. According to Vickers, most of Patten's catch "never left the region around Bedford; wither he carried it on horseback home to his farm, or he used

FIGURE 3. *Shad Fishing in the James River, Opposite Richmond,* from a sketch by W. L. Sheppard, in *Harper's Weekly,* May 9, 1874. A group of men are dragging a seine back toward shore. Seine fishing in the nineteenth century would have required the level of cooperative labor detailed in this image. Special Collections and Archives, Virginia Commonwealth University Libraries, Richmond, VA.

it to settle debts with people he knew."[32] Colonial and early national roads were sparse and too rough for the efficient transportation of goods. Though navigable rivers sometimes connected farmers to the marketplace, vessels that carried agricultural goods from inland sites of production to coastal centers of commerce were small and in operation only when conditions permitted.[33] As a result, market pressures did not entirely subsume these agricultural communities. Environmental constraints and the economic culture of the countryside discouraged unmitigated commercial exploitation, placing limits on the economic scope of preindustrial river fishing, despite the push to maximize the catch with large nets and cooperative operations.

The pursuit of a seasonally limited resource often led farmer-fishermen to combine efforts, increase efficiency, and harvest large numbers of fish, an approach that was not necessarily ecologically benign. The business of securing a comfortable subsistence led Patten and Robbins to fish beyond their immediate household needs.[34] However, before the turn of the nineteenth century, most farmer-fishermen fished only part-time during seasonal migrations. There was additional work to be done that precluded a singular focus on river fishing. On June 27, 1767, Patten wrote that "James Patterson and I catched a salmon between break of day and sun rise that weighed 12 lbs . . . and I ploughed with my own oxen and horse in the field above the barn and I went to the brook in the afternoon."[35] Traveling to and from Amoskeag Falls, back to his planting fields, and then a nearby stream, Patten balanced his fishing activity with agricultural work in Bedford.

Sylvester Judd's 1863 oral history of the South Hadley Falls shad fishery on the Connecticut River paints a similar picture. Judd characterized the majority of shad fishermen as "industrious famers . . . plainly dressed according to their business." Most stayed "one day or longer," and "after leaving the falls, they wound over the hills and plains with bags of shad in every direction."[36] J. W. Meader's 1869 natural history of the Merrimack River points to the seasonal abundance that brought fishermen to Amoskeag Falls who then "drew largely from them for the subsistence of their families, and though so many, there were fish for all comers." He continues, "One man had equal rights with another; the rule which secured the rights of each being tacitly understood and generally respected."[37] Despite a touch of nostalgia for a rural culture that preceded the emergence of industrial manufacturing, Judd and Meader emphasize the dominant values of traditional subsistence culture and republican equality that informed entry into the river commons.

Moreover, according to Sylvester Judd, fishermen gathered not only as enterprising yeomen but also "indulged in plays and trials of skill. Where there were so many men, and rum was plenty, there was of course much noise, bustle and confusion."[38] Farmer-fishermen imbibed spirits, cooked shad, competed in games, and generally enjoyed the fellowship of neighbors and friends after a long, difficult winter. Informal fishing companies served a parallel social function. Matthew Patten notes that his company helped forge new friendships and squash old rivalries. In 1773, the company accepted Major Goffe as a partner with the stipulation that he pay Patten a restitution of thirty pounds of salmon, "to make up my damage I sustained by his bad conduct these two years past." In all of these examples, river fishing provided a forum to reestablish connections, overcome old animosities, and simply enjoy one another's company. On May 6, 1773, Patten wrote that eleven members of his company, likely fishing with a seine, "began to draw a little before sunset and drawed till nine or ten o'clock and catched 67 shad in all we had 6 a piece," and all while sharing three quarts of rum.[39]

The Robbins and Patten documents illustrate a multi-occupational and seasonally variable economic culture that encouraged the intermingling of farming and fishing in the New England countryside. Harry L. Watson first introduced the concept of the backcountry farmer-fisherman in his study of class conflict in the antebellum South. Elite slave-owning planters, situated along fertile coastal plains, built milldams and monopolized seasonal fish runs at the expense of subsistence farmers further upriver. The economic power and legislative influence of the southern planter class served to "guarantee the monopolies of planter-fishermen and other representatives of commercial privilege." In the absence of these pronounced socioeconomic divisions in the preindustrial Northeast, rural farmer-fishermen consolidated even more support for equal rights to the river commons against the advancing "culture of the marketplace." Farming and fishing drew on the same values that informed New England's rural subsistence economies, what Christopher Clark describes as a "desire to acquire and hold onto the means of controlling their own effort and resources." James A. Henretta has emphasized the power of "lineal values" in which preindustrial farmers focused their efforts on passing along an inheritance "rooted in the land and in the equipment and labor needed to farm it."[40] Consistent with these interpretations, farmer-fishermen leveraged this lineal perspective when calling for the preservation of freshwater fisheries. In 1803, farmer-fishermen from Bradford, Massachusetts, complained that overfishing would "prove very injurious to the present and future generations."[41] Objecting to a dam proposal made two years earlier, townsfolk from Bradford recognized the "very great advantage that the present generation, and much more that of posterity, may derive by the passages for the fish that usually go up Merrimack River being kept clear."[42] This lineal ideology, coupled with the absence of a strong market orientation as evinced in the Robbins and Patten documents, opened space for the development and maintenance of sustainable relationships with freshwater environments. These petitioners expressed concern not just for their own immediate interests but also for the future survival of a critical natural resource.

Most farmer-fishermen in New England approached this resource as seasonal opportunists who returned to their fields after the dramatic runs of migratory fish had come to an end. They traded surplus fish within their local communities, salted or picked some in barrels for leaner times, and sold modest quantities for cash. They expressed concern that the destructive effects of overfishing would undermine their way of life. Freshwater fishing had long served these essential functions for Robbins, Patten, and the farmer-fishermen from Bradford. Further north, in the hill country of Vermont, river fishing also paved the way for Euro-American settlement and offered an important backstop against the distress and suffering of periodic crop failures or resource scarcity.

Fishing New England's Northern Frontier

Around the turn of the nineteenth century, northern New England attracted many an adventurous Yankee in search of socioeconomic mobility and independent yeomen status. With growing populations in the long-settled regions of southern New England, the hill country to the north represented a viable alternative for relieving these demographic pressures. As the lower Connecticut River valley filled with bustling towns and established communities, the northern reaches of the valley attracted greater interest. Between 1790 and 1830, the population of Vermont doubled, with growth in New Hampshire only slightly behind.[43] As these settlers of New England's frontier left the relative comforts of home and marched north toward unknown circumstances, they most certainly found great assistance in the region's forests, rivers, and streams.

During the mid-nineteenth century, Abby Maria Hemenway edited and compiled an impressive five-volume collection of local histories titled the *Vermont Historical Gazetteer*. A teacher, poet, and historian, Hemenway's project stressed the preservation of oral and written traditions that comprised the history of Euro-American settlement in Vermont. A collaborative project, Hemenway invited local authors from distant corners of the state to share their accounts of social, economic, and political development. Within many of these town and county histories, authors also detailed interactions and relationships with local environments. Thomas Waterman's account of the founding of Johnson, Vermont, pays special attention to the Lamoille River as a source of sustenance and support. Waterman relates that in 1784, Samuel Eaton, a veteran of both the French and Indian and Revolutionary Wars, journeyed seventy miles west from New Hampshire, cutting through dense wilderness and sometimes following "marked trees which he and his companions had previously glazed, while on scouting parties in the French war, and Revolutionary service." Searching for a prime location to homestead, Eaton scouted the forested landscape between Lake Champlain and the Connecticut River and ultimately chose a tract "on the right bank of the River Lamoille, on a beautiful bow of alluvial flats." Eaton's homestead occupied an ecosystem rich in biodiversity that would sustain his family until he could clear enough land to bring in the first harvest. Second-generation residents of Johnson told Waterman that these early settlers depended on "the forest and streams for subsistence. Moose and other native animals ranged upon the hills and mountains, and shoals of fish occupied the streams, and waters of the valleys; all of which afforded them a comfortable repast."[44] These local histories often stress not only the physical hardship that accompanied carving farm settlements out of the wilderness but also the environments and resources that afforded Eaton and other settlers a chance to clear the forest for planting fields. The second volume of Hemenway's *Vermont Historical Gazetteer* included selections from Seth Hubbell's 1826 account of founding the town of Wolcott. In 1789, Hubbell journeyed north from Norwalk, Connecticut, to Vermont's Lamoille River valley. Hubbell found the process of homesteading incredibly arduous. He managed to clear only two acres of land in the first year, writing: "My work was chiefly by the river. When too faint to labor, for want of food, I used to take a fish from the river, broil it on the coals, and eat it without bread or salt, and then to my work again. This was my common practice the first year till harvest."[45] In both examples, river fishing helped sustain first-generation Euro-American settlement in the hill country frontier. Rivers, as well as access to freshwater fisheries, helped newly arrived settlers remake the wilderness into a working pastoral landscape.

Fishing not only sustained new settlements along New England's northern frontier but also helped riparian communities overcome periodic crop failures and, on rare occasions, natural disasters. One particularly episode in the region's environmental history stands out in highlighting the importance of a multi-occupational and seasonally diverse resource base. Throughout the year of 1816, New England's climate experienced wild fluctuations and later earned the undesirable title of the "year without a summer," also characterized at the time as "eighteen hundred and froze to death." Persistent drought plagued the region, snow and frost appeared throughout the spring and summer, crops failed or lagged behind normal harvest predications, and, true to the famous phrase, harsh cold snaps struck without warning. A particularly terrible cold spell in June brought nearly a foot of snow to many parts of New England. Eighteen inches of snow fell in Cabot, Vermont. Erratic periods of cold coupled with an extended drought brought widespread hardship for the region's farm communities. Crop failures led to food shortages, which fueled hoarding and soaring prices for cereal grains. The root cause of such dramatic aberrations in average weather patterns lay with volcanic eruptions halfway around the world. Beginning on April 7, 1815, Mount Tambora on Sumbawa island, Indonesia, exploded with the reverberations reportedly heard one thousand miles from the site of the blast. Ten thousand people perished as the mountain shrunk over four thousand feet from the violent force of the blast. Volcanic dust spread throughout Earth's atmosphere and contributed to a decrease in average temperatures across the globe. Although other factors played a role in the dramatic shifts in global climate in 1816, experts still argue that this single event drove the overall cooling trend.[46]

New Englanders adapted the best they could to these dire circumstances. The town of Swanton, Vermont, in the northwestern portion of the state bordering Lake Champlain, found salvation in the nearby Missisquoi River. Absent the year's typical agricultural harvest, "the inhabitants of Swanton suffered in common with whole country for the necessaries of life . . . yet fish were plenty." In the fourth volume of her historical gazetteer, Hemenway featured the testimony of townsfolk who lived through these harrowing circumstances: "Probably never before or since were there so many fish caught in the river in the same length of time as in those two scarce years. There were probably not less than ten fishing-grounds between the Falls and the Lake, where large seines were drawn and kept in continual operation day and night during the fishing season. People in large numbers would come from the eastern towns with articles of barter of almost every description, to exchange for fish, prominent among which was maple sugar. This for fish was considered legal tender."[47]

During the "year without a summer," fishing took on a life-and-death character for the residents of the Champlain Valley. This account demonstrates the power of river fishing to help rural communities overcome these environmental calamities, as well as the personal exchanges that defined rural, preindustrial economic relations. Those who wished to secure a portion of the year's catch descended on these informal, but well-known, locales to enjoy an element of the seasonal diet not as easily acquired within their immediate environment. According to this account, folks traveled significant distances to participate in what could be described as seasonal fish markets. Fishermen flocked to these sites owing to the ease and predictability of a strong catch. The fishing grounds between the falls and Lake Champlain described above must have occupied a special place within the culture and economy of northwestern Vermont. Dragging nets between the falls and Lake Champlain on the Missisquoi River, fishers supplied enough protein to help the northern Champlain valley survive this time of scarcity, and luckily for these inhabitants, it seemed fish stocks were large enough to satisfy basic needs.

The storyteller claimed that never before had so many fish been taken from the river, which certainly is a function of the continuous operation of seines during 1816. These special circumstances necessitated such an intensive reaction, but absent this particular episode, farmer-fishermen would have resisted such a shortsighted decision to fish so heavily in this one location. Many fishermen recognized that too many fish pulled from the water could put the next season's catch in jeopardy. In fact, the typical New England diet tended to draw on a rather diverse array of foodstuffs that may have encouraged a more moderate consumption of fish in other years. Historian Sarah F. McMahon describes the seasonal nature of the New England diet: "The autumn harvest offered the most abundant supply of a variety of foods. From this perspective, the seasonal round of the diet becomes a process of depleting the stores of preserved food during the winter and spring, and of gradually adding new foods in the late spring and summer until harvest comes around again."[48] Swanton residents further elaborated on the place of fish within the local dietary regimen. They claimed that Vermonters traveled from the east each spring to trade for fish so as to "lengthen out the pork." They considered fish as "convenient to have in the house" during the spring season precisely for this purpose—to ration the remains of dwindling salt pork and beef and inject some much-needed variety into the diet. After a winter of salted meat, fresh fish must have been a welcome change of pace. And yet, rural communities would grow weary of an exclusive diet of fish. Most households alternated between pork, beef, and fish throughout the spring and summer until lambs were ready for slaughter. One Yankee saying reinforces the seasonal variation in the regional diet: "We hope meat will last till fish comes, and fish will last until meat comes."[49]

Swanton townsfolk remembered that during the "year without a summer," one well-known farmer journeyed to the Missisquoi fishery at his usual time, "with a load of barter to exchange for fish." He approached a familiar fishing shanty, where an acquaintance engaged him in the typical pleasantries: "Well, friend, how are ye, down again to get a few fish to lengthen out the pork? Eh?' The man replied, 'Not this time to lengthen out pork, but to lengthen out life."[50] While farmers might typically celebrate the arrival of spring and the passing of winter, the 1816 fishing season was much more a matter of life and death for the inhabitants of the northern Champlain valley. With stores of salt pork nearly used up by the end of the winter, New Englanders welcomed the opportunity to either catch or trade for fish, but this year was special. Salt pork and beef were typically expected to last through winter and at best into late spring.[51] However, the cold snaps of 1816 killed nearly the entire crop of hay, which led to shortages of feed for pigs, cattle, and sheep. Livestock perished due to the limited availability of feed.[52] Salt pork and beef likely did not last as long as expected under these difficult conditions. During this year, spring fishing stepped in when other parts of the New England diet failed to sustain inland agricultural communities.

Elias Robbins, Matthew Patten, and settlers of the hill country of northern New England embraced seasonal fishing as a fundamental component of their economic existence. River fishing fit seamlessly into the multi-occupational patterns of agricultural life, assisted the expansion of upland frontier settlement, helped communities weather unexpected environmental challenges as during 1816, and supported New England's yeomanry in the quest for economic independence. As enterprising yeomen or aspiring to independent status, farmer-fishermen shaped the world around them to meet their household needs and provide similar opportunities for their children. They transformed the natural world in accordance with the pastoral ideals of their European colonial predecessors, which sometimes placed demands on the natural world that the environment could not always accommodate. Yet these farmer-fishermen existed within a complex and sometimes contradictory world of economic motivations that, according to Henretta, prioritized the "yearly subsistence and the long-run financial security of the family unit . . . enmeshed also in a web of social relationships and cultural expectations that inhibited the free play of market forces."[53] This web of relations and expectations extended to the river commons, where fishermen turned to state authorities to criminalize destructive overfishing. The rural economy may have imposed limits on the commercial exploitation of river fisheries, but not everyone subscribed to this worldview.

Capturing the Commons

Along the Atlantic Seaboard and extending into the interior as far as large navigable rivers would allow, merchants, farmers, and fishermen inhabited an economic culture far more invested in market production. By 1700, the region's urban seaports functioned as vibrant centers of intercolonial and transatlantic trade. New England's merchant class supplied their colonial neighbors in British North America, as well as the slave plantations of the West Indies, with fish, timber, and farm produce.[54] They acquired and marketed manufactured goods from Great Britain and invested their profits in shipyards, offshore fishing fleets, and rum distilleries. Colonial and early national merchants, particularly in Rhode Island, also played an outsized role in capitalizing the transatlantic slave trade and marketing undesirable "trash fish," considered unfit for European markets, to their counterparts in the West Indies to feed slaves from Africa.[55] River fish never rose to the prominence of such staple commodities as cod, furs, timber, or rum in the larger colonial economy, but after the Revolution, while merchants renewed their transatlantic trade networks, they also looked for commercial opportunities closer to home.

Thomas Hazard, much like Matthew Patten, was a prominent figure in his community of South Kingstown, Rhode Island, and kept a detailed record of his daily life from 1778 until 1840. Known as "Nailer Tom," Hazard owned a small shop where he manufactured nails.[56] He also raised animals and grew corn and rye on a stretch of land on the Saugatucket River.[57] A respected member of his Quaker community, Hazard enjoyed a reputation for diligent record keeping and frequently acted as an arbitrator in local conflicts and formal legal disputes. In one case, authorities called on Hazard to testify in a dispute over fishing rights when a local fisherman set a seine that obstructed alewives from passing between two ponds.[58] Hazard's community looked to him as a steadfast source of expertise in a variety of town proceedings, but Hazard's background in the alewife fishery prepared him to speak as an authority on these matters. After the death of his wife in 1818, Hazard formed a fishing company with his son and a few friends. For the next three years, Hazard and his partners seined for alewives in nearby ponds, ecosystems that offered prime spawning habitat for this particular native anadromous species. Ironically, at the very same time Hazard fished the streams and ponds of South Kingstown, he also witnessed firsthand the technological transformations that would later represent the final nail in the coffin for New England's anadromous species: the manufacturing revolution brought about by the introduction of mechanized factory systems.[59]

FIGURE 4. "After a drawing by J. Milbert, Pawtucket Falls," 1825, Pawtucket, RI. Lithograph. RHi X3 1425. Note the Slater Mills across the river, above the falls, and the men fishing from the rocks in the foreground. Courtesy of the Rhode Island Historical Society, Providence, RI.

Unlike Patten and Robbins, Hazard's relationship with his local alewife fishery was more geared toward exchange in coastal markets. Hazard's proximity to Narragansett Bay allowed easy access to population centers with significant demand for salted and smoked fish. Most of Hazard's catch was sold for household consumption. Hazard noted his diligent work in the "fish house," smoking thousands of fish in preparation for sale in Providence and Newport. Without a doubt, Hazard's fish company ran an efficient operation. In May 1818, one of Hazard's partners, Joseph Taylor, set out with a cartload of 3,000 fish, bound for Providence. The very next month, Taylor again traveled to Providence, bringing 4,000 smoked fish, while Hazard busied himself salting an additional 1,600 fish. Hazard's company fished primarily with a seine operated by his son Benjamin and three friends. With large nets and the assistance of four strong men, Hazard's company consistently recorded impressive numbers. In June of the same year, the seine hauled in 1,056 fish, while Hazard kept himself busy in the fish house preparing the catch for market.[60] Hazard's description of spring and summer fishing reveals a well-organized and highly efficient operation, more so than that of Patten or Robbins, with a clear mission to harvest fish for sale in coastal markets.

These impressive numbers should not obscure the fact that Thomas Hazard did not fish exclusively for the market in Providence. In July 1818, he traveled to the home of George Hazard, the local blacksmith, and traded 120 smoked fish for a bushel of corn.[61] Hazard's fishing company labored in this fluid space between the marketplace and the subsistence values of the countryside. Christopher Clark argues that northern farms during the Early Republic could provide for daily subsistence while engaging markets to secure extra cash to pay taxes and purchase more land to meet the needs of the next generation.[62] Thomas Hazard supplied regional markets with smoked fish but also used the alewife fishery to interact with networks of barter and exchange. These fishing efforts both strengthened connections between community members and connected agricultural communities to regional market hubs, as demonstrated by Hazard's dealings around Narragansett Bay. Though Daniel Vickers argues that "the purpose of these companies was clearly to sweep the river of as many fish as possible," informal fishing partnerships paled in comparison with the commercial operations backed by wealthy merchants.[63] There is a legitimate basis for questioning the long-term sustainability of the type of fishing done by the Patten and Hazard companies, but in the end, it is important to recognize that they did not wholly depend on fishing to make ends meet. On the other hand, merchants like Justus Riley from Wethersfield, Connecticut, had little to no relationship with the subsistence values of the New England countryside.

The town of Wethersfield, Connecticut, has a long maritime tradition extending back to the colonial period. Despite its inland location, approximately fifty miles from the mouth of the Connecticut River at New Haven, sizable vessels could safely navigate several miles north of Wethersfield thanks to a broad and accommodating channel. Thus Wethersfield's riparian environment encouraged the growth of maritime trade. Shipbuilding and Atlantic commerce propelled the local economy and brought wealth and prestige to those engaged in these lucrative enterprises. From the colonial period through the Revolution, several industrious families enriched themselves through trade with the sugar colonies of the West Indies. Historians John J. McCusker and Russell R. Menard remark on these transatlantic connections: "The economies of the mainland and the islands were so tightly intertwined that full understanding of developments in one is impossible without an appreciation of developments in the other."[64] The West Indies provided major markets for New England's exports. Merchants and ship captains exported commodities scarcely available on a Caribbean island: lumber, horses, beef, pork, flour, and fish to feed African slaves. They returned to New England with stores of sugar, molasses, salt, and rum. The Riley family stood out among the elite members of Wethersfield society in this business.[65]

Justus Riley owned a fleet of sloops—single-mast vessels designed to transport cargo—and amassed significant wealth from the shipbuilding trade and transatlantic commerce. After commanding forty

men on the six-gun sloop *Hero* as a privateer during the Revolutionary War, Riley returned to the West Indies trade.[66] When Britain
closed its Caribbean colonies from American trade, Riley shifted
his operations to the French colony Saint-Domingue on the island
of Hispaniola. During one trip to Cape Francois in May 1792, Riley
delivered a cargo of horses, a schooner constructed in Wethersfield,
and other sundries for the price of twenty-nine-thousand pounds sterling.[67] Deeply engaged in the world of transatlantic trade, Riley also
looked for commercial opportunities around Long Island Sound.

Much like Matthew Patten and Thomas Hazard, Justus Riley
was deeply invested in New England's river fisheries. However,
Riley never put a line in the water, drew a seine, or constructed a
weir; there is no indication he did any fishing at all. Nevertheless,
several account books attest to his work in the propagation of an
interstate salmon trade. One such book from April 1783 shows that
Riley actively sought out salmon throughout the lower Connecticut
River valley and purchased as much as he possibly could. During the
month of April, he bought salmon from twenty-three different fishermen and shipped the catch across Long Island Sound to New York
City. One particular receipt shows that 740 pounds of salmon fetched
the price of fifteen pounds and eleven shillings. Riley typically com-
pensated fishermen with cash, but records show that he also paid
with goods like rum. After collecting more than three thousand
pounds of salmon, the sloop *Nancy* set out for New York City. Riley
settled up with the other three investors in this venture on June 8 and
took home half the profits from this particular voyage, which totaled
seventy-five pounds sterling. Although this amount pales in comparison with the proceeds from his West Indies ventures, Riley devoted
significant time and effort to the interstate salmon trade and kept a
steady pace, wheeling and dealing during the entire seasonal run of
Atlantic salmon.[68]

At the same time the sloop *Nancy* set sail for New York City, Justus
Riley worked with his relation Levi and other associates to stock two
additional ships with salmon. Levi Riley acquired more than three
thousand pounds of salmon and sailed the cargo across Long Island

Sound aboard the sloop *Black Joke*. He sold the fish in New York City and netted eighty-four pounds sterling. In May 1783, the next month after the sloops *Nancy* and *Black Joke* delivered their cargo, Justus Riley arranged for another trip to New York; this time, the sloop *Dolphin* carried some fifteen hundred pounds of salmon.[69] Altogether, Riley bankrolled three trips within a span of two months. Clearly, his interest in river fishing far surpassed that of Patten and Hazard in scope and scale. New England's coastal cities were built on a colonial tradition of transatlantic commerce and merchant capitalism, particularly as it related to cod, herring, mackerel, and other marine species. Wealthy merchants sought to integrate the salmon fishery into the world of Atlantic commerce and in the process extended financial incentives to inland farmer-fishermen to fish beyond the lineal subsistence values of family need and economic independence.

The account book for the sloop *Black Joke* details the purchase of salmon from four different fish companies and several individual fishermen. The fish companies often received goods in substitution for cash payments. Valuable items in the rural economy appear frequently in exchange for salmon: boat oars, salt, hemp rigging, and, again, rum. This speaks to a purposeful effort on behalf of the Riley family to encourage salmon fishermen to do business with them. When fishermen had more use for goods than cash, the Rileys accommodated with barter. In the Atlantic salmon fishery, elite merchants like the Rileys drew rural fishers into established economic networks with the promise of cash and valuable goods. Merchants outsourced the physical labor of fishing and offered financial incentive for fishers to look beyond the world of subsistence and neighborhood exchange.[70]

Thanks to the efforts of state authorities, the trend of increasing commercial activity is well documented. Throughout the nineteenth century, Massachusetts and Maine took an inventory of the number of barrels of pickled and smoked fish made available for sale throughout the region. A veritable army of state fish inspectors traveled between market hubs to ensure quality for retail consumption. From 1804 through 1805, markets in Boston, Newburyport, and

Portland combined to sell 184 barrels of pickled salmon. The sale of alewives expanded beyond the region's large cities into the hinterlands. Fish inspectors cataloged 5,000 barrels of alewives throughout Massachusetts and the Province of Maine during this one-year period.[71] Those numbers remained relatively steady almost a decade later, with 156 barrels of pickled salmon inspected throughout the commonwealth. Alewife landings hovered around the 3,000-barrel mark, while the year's shad run was particularly strong in Maine, with 1,300 barrels inspected.[72] These documents reveal a year-to-year fluctuation in the supply of river fish, as any number of environmental and human variables influenced the availability of the resource. However, they seem to confound the notion that the early nineteenth century brought a sharp decline in the salmon fishery. After Maine achieved statehood, fish continued to funnel into Boston at a steady rate. From 1823 to 1824, inspectors recorded 479 barrels of salmon brought into the city.[73] These nineteenth-century inventories of commercial salmon markets occurred at the same time fishermen pleaded with state authorities for tighter regulations to prevent overfishing.

Despite the intermingling of subsistence and market fishing, contentious issues of access, use, and distribution of a finite—and perhaps declining—resource brought farmers, fishermen, and merchants into direct conflict. Charles Buck, a fish inspector for the town of Hampden, Maine, laid out a potential source of conflict when he turned in his figures to state authorities in 1828: "I have made a distinction between fish brought in for country use and those caught in vessels belonging here and shipped to market." He noted that local fishermen preserved 1,310 barrels for "country use," while their commercially oriented counterparts shipped another 1,320 barrels to market centers in Portland and Boston.[74] This category of "country use" reflects the subsistence values of local consumption and community exchange. Fishermen enmeshed in the world of "country use" would have looked on with some suspicion as local resources were diverted to distant markets. Throughout the nineteenth century, farmer-fishermen advocated for regulations to constrain commercial fishing operations and contested the siphoning of fish resources from the countryside to urban spaces. Merchants worked to integrate small fishing companies into larger webs of market activity that transcended the boundaries of rural economic relations and the lineal subsistence values of farmer-fishermen. In the absence of the commercial limitations embedded within the rural economy, large merchants captured the commons for their own narrow economic interests. There is a tangible difference between hauling cartloads of preserved fish to a nearby city to satisfy local consumer demands and interstate shipments designed to enrich New England's merchant class—a group largely insulated from the physical experience of fishing and the ecological signals that foreshadowed decline and collapse. Merchants from Wethersfield were not alone in propagating this interstate salmon trade. Maine's first fish commissioner, Charles Atkins, described that "small vessels belonging in Southern New England used to visit several of the larger rivers annually and load with pickled shad and smoked salmon." High demand for salt shad meant that each season, "nearly all shipped out of the State." From 1842 to 1867, a single merchant firm in Boston purchased nearly the entire Maine shad harvest.[75]

The shift to intensive commercial entry into the commons was almost certainly unsustainable and not solely the product of merchant designs. This transformation required the participation of fishermen who understood how to get the most from their local freshwater ecosystems. Regarding the Connecticut River, historian John Cumbler argues that "by the early nineteenth century, shad fishermen were doing more than just feeding their families. The increasing numbers of poor and the shortage of arable land transformed shad fishing into a source of extra income, a way to gain extra cash to set up children in business or educate them, or just to buy needed goods. It also brought the market deeper into the lives of the valley's residents, farmers, and fishermen."[76] Merchants like Justus Riley opened a door that accelerated this transition toward commercial production and spelled the end of common resources. Given the forces that slowly shifted river fisheries into the landscape of profit and commerce, a rural economic culture that encouraged a seasonally variable subsistence offered more room for sustainable relationships with freshwater ecosystems.

This period witnessed the evolution of two distinct but interwoven economic systems with competing understandings of the proper relationship between fishermen, the natural resources on which they had come to rely, and commercial markets. Conflicts inevitably arose as a result of the conflicting goals with which fishermen approached this resource. These divergent economic worldviews shaped the way fishermen approached the commons, and it appears that such merchants as Justus Riley, along with the burgeoning class of commercial fishers who did business with them, represented a significant threat to the function of river fisheries within the traditional rural economy. His operation significantly diminished the number of fish available for communities on the northern peripheries of the Connecticut River, communities that lived alongside hundreds of miles of prime fishing grounds. In 1788, the towns of Northumberland and Lancaster—in the upper Connecticut River valley of New Hampshire— witnessed steady declines in the number of fish reaching northern spawning habitat and argued that their southern neighbors were to blame. As fewer and fewer fish appeared in these waters, particularly salmon and shad, communities fought to maintain access to a critical element of their subsistence economy.[77]

A multitude of demographic and economic pressures encouraged more intensive pursuit of anadromous fish. Demographic pressures also tested the strength of New England's soils. Many answered the call of western migration; those who stayed intensified their use of diminishing resources. New England's rural economy began to shift in the direction of specialization and commercialization.[78] Farmers experimented with more efficient techniques or abandoned farming altogether in favor of commercial fishing, providing a source of food for sale at the larger markets of Boston and bait for offshore fishing fleets.

Thus began the specialization of freshwater fisheries throughout New England, which challenged the remaining agricultural communities that still sought their share of this valuable resource. With a new market economy on the horizon, farmer-fishermen endeavored to maintain their right to take fish as a matter of equity and practical subsistence. With solid evidence of steady decreases in river fish, farmer-fishermen campaigned for regulations that would reverse the trend. They held to an ideology of open access and equitable distribution, a rights-based understanding of shared resources that shaped the history of inland fisheries management as communities of farmer-fishermen sought to protect a traditional relationship with the environment.

CHAPTER FOUR

COMPLICATIONS IN
THE COMMONS

In August 1811, James Cochran and dozens of his neighbors signed a petition addressed to the Massachusetts General Court. Frustrated, angry, and determined to expose what he considered serious transgressions by certain members of the fishing community, Cochran offered a scathing rebuke of illegal fishing at Pawtucket Falls on the lower Merrimack River, a stretch of water that would soon launch the manufacturing boom and transform this rural backwater into the factory town of Lowell, Massachusetts. A thirty-two-foot drop, Pawtucket Falls, as with all dramatic cataracts across the region, forced salmon and shad to school in large numbers below the falls as they struggled to ascend this natural obstacle. For this reason, pools immediately below and above the falls attracted a significant amount of fishing pressure. These natural features, combined with a keen ability to time seasonal fish runs, nearly guaranteed the opportunistic fisher a chance to capture more than his fair share. Cochran himself witnessed "seines, nets, pots and other machines" employed on days "without regard to, and in clear violation of law." Even worse, Cochran noted, offenders, to avoid penalties, "dress themselves in disguise, which prevents the fish wardens from knowing them and consequently from bringing prosecution against them."[1] For Cochran and his fellow petitioners, these fishermen had breached both the legal structures and informal arrangements meant to discourage overexploitation and promote an ideology of equal access. Cochran wrote the Massachusetts General Court to condemn and correct these flagrant violations of law and custom. Luckily for Cochran and his cohort, the Massachusetts General Court proved a responsive ally. As a result of his petition, in February 1812 the legislature passed a law to ensure the "preservation of salmon, shad, and alewives in their progress up the Merrimack River."[2]

The term "preservation" was no throwaway line. It first appeared in a 1741 colonial statute and in 1781 became a consistent refrain in colonial and early national fisheries legislation.[3] For decades, this call for preservation intersected with the recognition that, according to James Cochran and his fellow petitioners, river fishing held "great importance to the inhabitants of those town through which the said river and streams pass, and to the public at large."[4] Whether they identified it as a great importance, service, or public benefit, rural communities invested tremendous effort to protect this seasonal resource, and farmer-fishermen and legislators alike clung to this message of preservation as the guiding force for decades of inland fisheries regulation.

For farmer-fishermen like James Cochran, committed to conserving inland fish stocks for themselves and future generations, the pivotal concerns centered on the damaging effects of overfishing and obstructions to critical upstream spawning habitat. As market pressures increasingly gave rise to a class of full-time commercial freshwater fishermen, competition for this finite resource sparked a protracted debate over the boundaries surrounding appropriate and legitimate entry into the river commons.[5] Conservation-minded agrarian fishermen complained that seines, nets, weirs, and the like—in operation for the entire duration of anadromous fish migrations—swept too many fish from vulnerable locales, preventing some from passing upriver to reproduce.[6] The feeling that commercially oriented fishermen, merchants, and owners of small mills not only abused this resource but also deprived their fellow citizens of rights of access established in common law and custom, inspired a nearly century-long grassroots campaign to conserve river fisheries. Conflict arose as farmer-fishermen struggled to enforce community norms and fishing practices thought to produce a sustainable catch.

Overfishing, milldams, and other obstructions stood in the way of this quest for a functional equilibrium.

Despite the egalitarian principles, both codified in law and generally accepted, that supported relative open access to the river commons, overfishing remained the singular threat to preindustrial river fisheries. Commercial fishermen worked to take as many fish as a river could provide. On the Connecticut River, they worked cooperatively using boats, seines, and scoop nets to take as many as two to three thousand shad in one day.[7] Near the mouths of brooks or streams where anadromous fish typically ascended in search of spawning habitat, commercial fishermen used seines and other large nets to capture schools of fish, and the consistent use of this equipment, operating nearly uninterrupted throughout the spring and summer fishing season, undoubtedly had a deleterious effect on New England's river fisheries. Though these practices threatened to deplete and destroy migratory fish populations, commercial fishermen made a tidy profit funneling barrels of shad and salmon to retail markets in urban seaports. As long as money could be made, they continued to violate regulations designed to prevent the kind of widespread overfishing commonly seen at the turn of the century.

Because farmer-fishermen did not rely on this resource alone to secure their economic fortunes, they were more inclined to look toward the future health of river fisheries. The practices described in the diaries and account books of farmer-fishermen such as Matthew Patten and Elias Robbins reveal a strain of aggressive country fishing, but not a naked commercial ambition. They fished as part of a larger quest for economic independence—values spawned in the Revolution that became inseparable from the yeoman ideals that flourished during the Early Republic. The depth of this commitment was undone only by the arrival of industrial dams. Before this fundamental shift in the New England economy, farmer-fishermen prodded state legislators to act on their behalf and manage this resource for the benefit of all. They called for regulations to limit the time, location, and placement of potentially destructive fishing equipment and trusted these demands, translated into law, could preserve river fisheries.

The Moral Commons

Fishermen who overexploited this resource not only jeopardized future returns but also violated the customs and ethics of the river commons to which so many farmer-fishermen subscribed. Those who cautioned moderation and restraint operated with a surprising degree of local consensus. The decades-long legislative pursuit to tamp down on overfishing does seem to indicate broad agreement that monopolization by the few was unacceptable. If it had been otherwise, hundreds of state fisheries statutes, lacking broad support, would have been relegated to the legislative graveyard. In this sense, the majority of farmer-fishermen who either eschewed seines and weirs or favored a much more limited use thereof deserve far more credit as the ideological defenders of a river commons, a far-reaching natural philosophy embedded within decades of petitions and corresponding measures to protect inland fisheries.

Farmer-fishermen communicated these grievances through established legal channels and petitioned state legislatures in the hopes they would accept recommendations from those who best understood their local environments. These grassroots management proposals were supported by deep environmental understandings on the part of fishermen themselves, knowledge steeped in generations of working closely in the natural world. These local networks of environmental knowledge informed a series of petitions as farmer-fishermen demonstrated a commitment to preserving a more natural vision of freshwater ecosystems. This conservation vision pivoted on the necessity of free-flowing channels to allow anadromous fish the opportunity to complete their life cycle and spawn the next generation. They understood the consequences of upstream habitat loss and fought to keep rivers and streams open to anadromous migration.[8]

Farmer-fishermen leveraged their knowledge of anadromous migration and local fish stocks with the hope of influencing inland fisheries regulation through legislative petitions. Petitioning has long served as an outlet to express economic and political frustrations. After the Revolution, the citizenry took advantage of those channels to not only advocate for greater economic opportunity, as Ruth Bogin argues, but also to make "recurrent demands for specific measures to diminish the spread between the bottom and top economic levels and to protect weaker economic groups from exploitation."[9] Bogin's analysis of late nineteenth-century petitions underscores a moral economic vision in which the lower classes marshaled a defense of economic independence and expressed a keen awareness of the potential for exploitation by elite interests within and outside the government. On New England's northern periphery, this ideology often clashed with mercantile interests based further to the south in cities like Boston.

After the American Revolution, Maine's frontier settlers fought off the tenuous claims of wealthy land speculators who sought to evict the self-styled "Liberty Men" from wilderness land they cleared, planted, and made productive through their own labor. Derided as squatters and lawbreakers, these yeomen farmers defended the long-held belief that title rightly belonged to those who had "improved" wilderness land for agriculture.[10] Just as Maine's Liberty Men brought this moral dimension to discourses of land tenure and resource rights, New England's farmer-fishermen interpreted resources that sustained the rural economy in both moral and political terms. For farmer-fishermen, this was a question of the material well-being of New England's agricultural communities—a bareknuckle contest over the means to secure a livelihood. Fishermen wrote that those living on the margins of the Merrimack River "annually procure large quantities of those fish for the sustenance of their families," and in harvesting the seasonal fish runs, they could secure a "little cash to help the poor fishermen to pay his taxes."[11] Fishermen also noted that salmon, shad, and alewives, were "of great value . . . and the business of catching them is a great advantage to the poorer class of people."[12] Whereas the Liberty Men often turned to extralegal means to resolve disputes between these poor squatters and land agents under the employ of wealthy speculators, farmer-fishermen worked through established legal channels to protect a river commons defined by open and equal access and fair distribution of fish resources.

These petitioners attempted to voice what E. P. Thompson describes as a "traditional view of social norms and obligations . . . the proper economic function of several parties within the community."[13] Historically, when these outlets fail to release the valve of socioeconomic frustration, disillusioned subjects or citizens take to the streets in demonstrations of violent protest. In the case of Britain's eighteenth-century food riots, Thompson argues that "the men and the women in the crowd were informed by the belief that they were defending traditional rights or customs; and in general, that they were supported by the wider consensus of the community."[14] Though rarely rising to the level of direct action, riots, or extralegal violence, farmer-fishermen operated within a similar framework of community consensus. They understood themselves as defending the traditional rights and customs that supported the material well-being of New England's rural communities. Farmer-fishermen were supported by the broader community of agrarian country fishers as evidenced by the sheer number of petitions and community members willing to lend their signatures and align themselves with this evolving conservation ethic.

American political leaders, conscious of the more chaotic and violent expressions of political protest after the Revolution, kept open channels of communication between the citizenry and centers of political power. With civil unrest rampant throughout the Early Republic, evidenced by Shays' Rebellion and similar episodes of collective violence, many Americans were disillusioned with rising inequality, the shrinking opportunities to join the ranks of the yeomen class, and the feeling that the promises of the Revolution were slipping away. Landownership represented the most traveled road to realizing the goal of independent status, but a host of resources contributed to economic success in the countryside. It is then no surprise that farmer-fishermen drafted hundreds of petitions to protect river fisheries steeped in the language of republican equity, economic independence, and the public good.

These documents offer a glimpse into the ways fishermen understood their impact on freshwater fisheries and their relationship to the broader environment. In the face of commercial pressures, farmer-fishermen responded with a defense of the public's right to access this shared resource and endeavored to protect their right to harvest river fish as a matter of principle and practical subsistence. They consistently strove to maintain sustainable river fisheries, fighting to curb destructive fishing practices, remove obstructions, and force local mill owners to construct passages so that migratory fish could pass upriver. Fishermen sensitive to the consequences of overexploitation and wholesale alterations to river ecology appealed to the legislative bodies of their respective states, reporting that inland fish stocks were in serious decline. They argued that should weirs, seines, and other obstructions continue to restrict the free passage of fish upriver, future generations could not rely on strong seasonal returns. State authorities responded with a series of legislative remedies aimed at the preservation of salmon, shad, and alewives.[15]

Management on the Main Stem

Fishers, farmers, and petitioners from a variety of backgrounds with a common stake in river fisheries put their faith in regulation as an instrument of preservation, with enforcement by duly sworn fish wardens, justices of the peace, and town selectmen. In colonial New Hampshire, with a fisheries statute set to expire after a term of three years, fishermen wrote to the legislature to continue these regulations. The fishermen got their wish when in 1767, the New Hampshire General Court renewed the law for an additional five years thanks in part to local testimony that the "aforesaid has been found by experience, to be serviceable for the increase of the fish in the River of Merrimack."[16] Thirty years later in 1797, New Hampshire petitioners continued to argue that fishermen "near the River Merrimac . . . have some knowledge of the benefits arising to this part of the Community, from Salmon, Shad, and Alewives taken in said River and the water falling thereinto." According to the petitioners, these community benefits, or "privileges," had "been abused by many who have, unduly, obstructed the passage of fish in the streams, whereby they have been much decreased for many years."[17] Farmer-fishermen leveraged their knowledge of local freshwater ecosystems, predicated on generations of experience, to advance an argument that regulation was necessary to preserve migratory species and protect public rights to access the river commons. Later that year, legislators answered with a law disallowing the use of seines and nets close to milldams, sluiceways, and the mouth of streams emptying into the Merrimack River.[18] State officials affirmed public fishing rights and worked with locals to curb overfishing and milldam obstructions that compromised the ecological balances sustaining these migrations.

Even in the midst of steady declines, fishermen held out hope that wise regulation and strict enforcement could break reoccurring cycles of exploitation and depletion. The petitioners from the Merrimack River valley in New Hampshire wrote that "although legislation in the year 1795 . . . had had a tendency to increase the fish, yet it is so deficient that it has not fully answered the salutary ends therein intended."[19] In 1801, on the lower Merrimack River, fishermen from Bradford, Massachusetts, similarly argued that some progress had been made: "We find from experience that the wise and wholesome laws for regulating the fishery on Merrimack River in the present partial execution of it has in some measure retrieved the almost exhausted run of fish."[20] Far from embracing the inevitability of collapse, farmer-fishermen clung to the notion that regulation, under the right circumstances, might reverse these declines and, according to petitioners in 1776 Derryfield, New Hampshire, perhaps even "induce the said fish to return and . . . increase in the future."[21] Rural communities and legislators emerged from the Revolution prepared to use colonial fisheries statutes, as well as the experiential knowledge of conservation-minded farmer-fishermen, as a foundation for new state-based conservation programs.

Around the same time as the Treaty of Paris in 1783, the Massachusetts "Senate and House of Representatives in General Court assembled," met to repeal colonial legislation and institute their own "act to regulate the catching of salmon, shad, and alewives, and to remove and prevent obstructions in Merrimack River, and in the other rivers and streams running into the same, within this Commonwealth." The preamble clearly spells out the intent of the act, declaring that "salmon, shad, and alewives . . . have been of great service to the inhabitants of this Commonwealth, as well as those of New Hampshire." The act asserted that the "constant fishing with seines, weirs, nets, pots, and erecting weirs and other incumbrances" prevented upstream passage for these anadromous species and acknowledged the "great danger of their being totally destroyed."[22] As a result, the General Court passed a statute that limited fishing to Tuesday, Wednesday, and Thursday and labeled weirs and "other incumbrances now erected" that stopped the passage of fish upriver a "common nuisance" that should be "pulled down and demolished." Furthermore, lawmakers limited the size and placement of seines and other large nets, banned the use of seines near the entrance of streams that led to spawning habitat, and instituted a system by which towns bordering the river would each elect "at least four suitable and fit persons" to see that "this act, and the acts for keeping open sluice ways in dams, be duly observed."[23] In an effort to harmonize their approach with Massachusetts, in April 1784 the State of New Hampshire imposed similar restrictions that prohibited fishing the Merrimack River outside of Tuesday, Wednesday, and Thursday.[24] Two years later, New Hampshire lawmakers amended the statute to strengthen protections against milldams, weirs, and other obstructions to upstream passage.[25] This degree of interstate cooperation set an important precedent for large river systems throughout the region. Effective regulation would have to consider the entire watershed: streams, tributaries, headwaters, and estuary, regardless of state boundaries.

Lawmakers on both sides of the state border continued to experiment with qualitative use-limits related to the time, place, and manner one could fish the Merrimack River. Unable to wholly interrupt destructive fishing practices or milldam obstructions, the Massachusetts General Court again took up the issue in 1789, repealing the previous act and replacing it with one that, in its six pages, attempted to leave no stone unturned. The new law left many of the previous regulations in place but raised most penalties for violations and introduced the new office of fish warden. Elected by "every town in this Commonwealth bordering on the Merrimack river," at least four fish wardens were sworn to "pursue and execute the duties of their office," ensuring that dams maintained open sluice ways, informing against those who willfully defied the law, examining and measuring nets and seines, and removing all unapproved obstructions. Leaving no doubt that the fish warden now represented the primary mechanism of enforcement, the act imposed fines against towns that neglected to fill vacant positions and penalties for obstructing fish wardens in the execution of their duties. With the most vulnerable locations along the watershed in need of some measure of supervision, the General Court called on grand jurors, sheriffs, deputy sheriffs, and constables to aid wardens in the apprehension and prosecution of offenders.[26] Not to be outdone, from 1790 to 1795, New Hampshire issued three more acts to "prevent the destruction of salmon, shad, and alewives" in the Merrimack River that similarly instituted a new class of fish wardens for towns bordering the river.[27] During this period, both state legislatures experimented with the number of closed days, the placement of fishing gear within the ecosystem, the powers and responsibilities of elected officials to enforce regulations, and the best manner to safeguard free passage to inland spawning grounds, but it seems no one approach represented the silver bullet that would once and for all address the environmental impact of obstructions and overfishing.

This series of regulations, though an important first step toward the preservation of salmon, shad, and alewives, fell short of stated goals. In 1797, New Hampshire townsfolk continued to voice their frustration that fishermen and millers "obstructed the passage of fish in the streams." They praised the 1795 law, which in their estimation "has had a tendency to increase the fish" but was nevertheless "so deficient that it has not fully answered the salutary ends therein intended." They hoped to "enlarge the powers of the fish wardens . . . that they be empowered to command assistance (if necessary) in the execution of their office" and that "persons found dragging any net in the waters aforesaid at any time when fishing is prohibited by law . . . forfeit and pay the sum of 10 dollars" and forfeit their equipment.[28] For these petitioners, years of legislative remedies to enforce conservation measures failed to answer the ongoing problem of overexploitation. This concern has plagued shared resources for generations. As with any common-pool resource, stakeholders must confront the uncomfortable truth that, according to Elinor Ostrom, they "are sufficiently large that multiple actors can simultaneously use the resource system and efforts to exclude potential beneficiaries are costly."[29] For the next fifty years, petitioners and lawmakers would repeal, amend, and replace legislation governing the Merrimack River to meet the high cost, in time and effort, of excluding bad actors.[30] This process of experimentation played out within every large watershed throughout the region—a sustained effort to identify the correct balance of regulation and enforcement. Fishermen and state authorities initially prioritized the region's most significant watersheds, in terms of size and abundance of fish resources, as targets for reform and regulation. The Connecticut River, New England's largest river system, witnessed decades of concerted effort to preserve traditional river fishing as a component of the agricultural economy.

Conflict and Cooperation on the Connecticut

Turn-of-the-century fisheries statutes had to reconcile the desire to ensure equal opportunity to access the river commons with the reality of disproportionate commercial fishing pressure, particularly on large systems like the Connecticut River. Charles Atkins, Maine's first fish commissioner, described this tension at the center of "legislation on the river fisheries . . . first, the preservation of the supply of fish; second, the harmonizing of conflicting interests."[31] Because agrarian subsistence interests stood in opposition to the rising commercial presence in freshwater ecosystems, neither goal would be so easily accomplished. These conflicting interests played out along market and population centers near the coastline, but also near natural obstructions that attracted intense and often unsustainable fishing pressure. On the Connecticut River, migratory fish concentrated in large numbers below waterfalls seeking to pass upriver, which attracted large numbers of subsistence and commercially oriented fishermen as demonstrated by Sylvester Judd's oral history of South Hadley Falls in western Massachusetts.[32] The ease of fishing at these locations, as well as the inclination toward cooperative efforts, presented complications for sustainable river fisheries. In fact, these environments encouraged equality of access but also undermined fishing rights for farmer-fishermen further upriver, the classic commons dilemma. Charles Atkins's analysis proves instructive for understanding how petitioners and lawmakers worked to address this challenge, with "special laws instituting town fisheries in which all citizens should have an interest," thus "restrict[ing] certain methods for the benefit of others."[33] This maxim guided every legislative program to protect river fisheries from commercial abuse, particularly for large watersheds like the Connecticut River.

In May 1788, the towns of Lancaster, Stratford, Dartmouth, and Northumberland, situated within the upper Connecticut River valley of New Hampshire, petitioned the legislature to restrict fishers whose actions threatened their ability to access the river commons. The petition identified those who "have combined together" on the "Great Falls" on the Connecticut River, particularly Bellows Falls, and installed large nets across the entire width of the river at the height of fishing season, "which stop all the salmon." They complained that commercially oriented fishers "make it their business" to set their equipment directly at the foot of the rapids or otherwise "in the only places where the salmon can pass or get up said falls, as there is but one or two places that they can any way pass."[34] Even though more than one hundred miles separates Bellows Falls from these north country towns, fishermen understood that salmon obstructed in such a fashion so far from spawning grounds—cold, gravel-bed streams within the mountain headwaters—would not only undermine fishing opportunities for their frontier communities but also jeopardize the entire Connecticut River fishery. They demanded protections to uphold an important though rudimentary ecological vision: to prevent so many fish from reaching their spawning habitat to reproduce the next generation would ultimately assure the disappearance of this resource altogether. A succession of natural obstructions on the Connecticut—South Hadley and Bellows Falls being just two fishing locations featured prominently in the historical record—attracted a tremendous amount of localized fishing pressure. Aggressive fishers pulled incredible numbers of fish from the water on the journey to their northern spawning grounds. In addition, merchants such as Justus Riley compounded the problem of depletion for these residents of the north country by integrating local fishers into interstate salmon markets further south in the tidewater.[35] Riley purchased nearly every available salmon from fishermen in the greater Wethersfield region, and fish that managed to evade those in Riley's employ faced a labyrinth of large nets and other obstructions on their journey to northern spawning grounds. Under these circumstances, it was remarkable any fish at all could succeed in navigating against the current as far north as Lancaster or Northumberland. With so many fishermen harvesting from the Connecticut along its southern reaches, it is no surprise residents of the upper Connecticut River valley turned to the state legislature to address this imbalance.

Perhaps most revealing, the north country petitioners clearly differentiated between these strains of commercial exploitation and their preferred method of spearing salmon, a fishing strategy that originated with the Pocumtuc and Penacook peoples of the Connecticut and Merrimack Rivers. The fishermen explained that, historically, "persons among us . . . used to stabb with their spears 18 or, 20 salmon in a night," but there "now scarcely is a salmon to catch."[36] Seines, large nets, and "other obstructions" substantially increased landings for fishermen at Bellows Falls, but at the expense of the petitioners further north. Theodore Lyman and Alfred A. Reed's 1866 report on the decline of freshwater fisheries in Massachusetts helps illustrate the relative impacts between these two fishing practices. They spoke with Charles Ramsay, who fished with these large nets on the lower Merrimack River in Amesbury, Massachusetts. Lyman and Reed write that "not only shad, but salmon, were at that time very plenty. It was customary to get, with a ninety-yard seine, from sixty to one hundred salmon a day. Such a day's fishing would now be worth from three hundred to five hundred dollars. He once took eighteen salmon at a single haul and has seen twenty-four taken."[37] This description speaks to the influence of the marketplace on fishing activity in the tidewater. Ramsay would have had easy access to markets, turning his catch into cash in the next town over, Newburyport.

Seine fishing offered opportunities to boost efficiency and increase the catch, however the north country farmer-fishermen identified specific circumstances in which the strategic placement of seines and other large nets on the "Great Falls" of the Connecticut River "in all Probability will stop every salmon as they have almost done it in years past."[38] These frontier communities had come to rely on the abundant spring fishing season and could not abide those further to the south that undermined an essential component of their economic life. Interestingly, Lyman and Reed's 1866 report also addressed the importance of this resource for subsistence communities on the northern frontier and highlights the spatial dimensions of these types of resource fights. They collected testimony of the salmon fishery in the Pemigewasset River, a channel that, in conjunction with the Winnipesaukee, forms the Merrimack River in northern New Hampshire: "Mr. Barron, of Woodstock, New Hampshire, aged seventy-eight, recollects when they came up in vast numbers. It was then the habit for each family that lived near the stream to lay in some four barrels of salted salmon, which would be equal to about one hundred fish. The 'eddies,' or pools, where the salmon loved to lie, were all numbered, and the fishermen had certain customs and rights in them. It is in tradition that one capable old lady was wont to spear the fat fish with a pitchfork, to feed her working men withal."[39]

This account adds an important dimension to the socioeconomic underpinnings of fishing in northern New England. These communities caught and stored salmon as an important component of the region's seasonal diet. Those four barrels of preserved fish, approximately one hundred fish for the season, represented a lifeline for agricultural communities on the northern frontier and yet a paltry figure compared with Ramsay's testimony of one hundred salmon in a single day. Residents of Woodstock, New Hampshire, just like those of the upper Connecticut River valley in Lancaster, Dartmouth, or Northumberland, had good reason to fear that a Justus Riley or Charles Ramsay—working large nets throughout the entire seasonal fish run—could deny those on the northern frontier their fair chance to access the river commons. These families on the margins of the Pemigewasset River entered the fishery to store away a much-needed surplus that supported the household economy. Deep understandings of local fishing practices helped ensure that households could capitalize on this seasonal resource.

As water flows past an obstruction, perhaps a rock or ledge, an eddy can form up against the river bank or immediately downstream of the obstruction. As a result, water inside an eddy flows in the opposite direction of the current. This can sometimes produce a current that swirls violently against the normal flow of the river. Alternatively, eddies can generate stretches of calm water that attract fish weary from the difficult upstream migration. For this reason, Woodstock fishers identified these eddies as particularly fruitful for salmon fishing and devised local arrangements to protect these spaces as a shared resource. Townsfolk allocated salmon pools, or eddies, to those living near streams and brooks within the Pemigewasset watershed so that every household could collect their four barrels. These informal arrangements reflected local networks of environmental knowledge as it related to the behavior of salmon and other anadromous species. As anadromous fish battled their way upstream, local farmer-fishermen adapted their fishing practices according to these environmental understandings. As a result, some fishing places acquired reputations that caught the attention of the entire Woodstock community, so much so that townsfolk instituted informal rules and norms to govern the local salmon fishery.

Set against the plight of New Hampshire's frontier fishermen, this account clearly separates fishermen into distinct categories based on the type of fishing gear, choices of fishing location, and overall motivation: subsistence or commerce. The practice of spearing fish was a laborious, time-consuming effort geared toward household use and family consumption. The use of seines, nets, weirs, and the like increased the catch substantially and, in this case, perhaps at the expense of upriver communities seeking a fair share. Frontier communities relied on the abundant spring fishing season and could not abide those who undermined such a valued article of their economic life. Petitioners understood well the difficulties of forging a subsistence economy within locales that possessed less-than-desirable soils. They described their condition as "we that are settling and cultivating through every hardship the Newlands, and at a great distance from the sea coasts." In northern New Hampshire, known more for its rugged, mountainous terrain than fertile soils, frontier communities had come to rely on hunting, fishing, and foraging to secure a comfortable independence. Situated hundreds of miles from the coast, anadromous fish nevertheless followed their biological instincts, against tremendous odds, to reach the cold streams near the White Mountains. For these fishermen, such remarkable journeys were imbued with spiritual meaning, arguing that they should not "be deprived of what the all wise being has in his wisdom provided for us."[40] Farmer-fishermen interpreted these dramatic migrations as gifts from a benevolent creator—not altogether dissimilar from the animistic worldviews that informed Indigenous relationships with the natural world.

New Hampshire's legislature responded to these requests, though perhaps not exactly as townsfolk from the north country would have hoped, with limits to the number of days one could access the fishery and a prohibition on nets longer than twenty rods (110 yards).[41] Again, with petitions continuing to flood the New Hampshire legislature, subsequent statutes offered subtle tweaks in the number of days allowed for open fishing, the size of nets and seines, and the proper function of fish wardens.[42] Petitioners from northern New Hampshire were not alone in their belief that further regulation was required to address overexploitation within these vulnerable locations.

In June 1788, fishermen from Hampshire County in western Massachusetts, nearly 200 miles south along the Connecticut River from the New Hampshire petitioners, echoed the concerns of their northern neighbors that seines and other large nets swept far too many fish from the river. Hampshire County fishermen witnessed firsthand the devastating effects of overfishing at South Hadley Falls. Legislation included testimony that shad and salmon had "decreased for a number of years past and that there is a great danger that this fishery in said river will be destroyed."[43] Something had to be done to alleviate the overwhelming fishing pressure associated with waterfalls and other well-known, productive fishing locations. Petitioners decried the choice of seines and other large nets as the preferred fishing equipment of the commercially minded fishermen, so limits on the size and placement of this gear seemed a good place to start. The Massachusetts General Court issued a statute designed to address these concerns. The law applied more qualitative use limits that opened the fishery four days a week and established a closed zone one mile below South Hadley Falls during the fishing season, with the exception of Tuesdays and Wednesdays.[44] While lawmakers worked to reduce the impact of seines and large nets within vulnerable locations, many fishermen felt unfairly targeted given that inland fishing regulations could vary widely from state to state. Massachusetts fishermen did not want to lose out while aggressive fishermen further south continued to draw their seines uninterrupted.[45]

Anadromous fish care little for interstate borders. Their migrations through the region's large river systems traverse multiple states across the Northeast. For the Connecticut River, some migratory fish may pass through the river's namesake, Massachusetts, New Hampshire, and Vermont before they finally reach northern spawning habitat. To reconcile state regulations, many fishermen called for interstate cooperation to arrive at some uniform conservation program for the entire watershed.[46] Hampshire County fishermen felt unfairly subjected to strict controls and limitations, whereas

Connecticut residents further south could exploit this resource free from state intrusion, all the while diminishing the catch for fishermen upriver. They desired that the same regulations apply to the entire Connecticut watershed. This seemed a reasonable proposition, but as with most distribution conflicts, interstate cooperation was no easy task as each state aimed to satisfy the needs of their respective constituents.

Surely aware of the potential for conflict, Massachusetts sent a committee to Connecticut to coordinate and "devise the most effectual means of preserving fish in said river."[47] The meetings must have been surprisingly productive as both sides agreed on a few basic principles. The backbone of this new joint understanding between Massachusetts and Connecticut rested on the recognition that seines and large nets could not be allowed for the entire duration of the migratory fishing season. As a result, from March 15 to June 15, both states restricted seine fishing to four days a week. After June, they mandated the removal of seines and other "machines" constructed to catch fish, with fish free to pass upriver for the remainder of the migratory fishing season. In addition, the states prohibited the use of seines longer than twenty rods.[48] Since salmon migrations can carry into the early fall, this law allowed ample time for salmon and shad to pass upriver free from obstructions, presumably with the exception of less-invasive gear: a dip net, spear, or hook and line. The governor of Massachusetts, John Hancock, sent the coordinated statute to New Hampshire and Vermont for further consideration.[49]

New Hampshire policymakers obliged this emerging interstate coalition, backing this specific regulation, and passed a statute that largely echoed the call for uniformity in main-stem river management. The legislature allowed seines and other "machines" used to catch fish between March 15 and July 15.[50] With salmon and shad near the northern reaches of their migrations, fishermen and policymakers from northern New England lobbied for an additional month of open fishing on the back end of the season. Their counterparts to the south apparently acquiesced on this point, and a tentative consensus was reached.

Prior to this search for regulatory consensus, Vermont took an equally hard-line position with seines and large nets. In 1787, still an independent republic, Vermont passed a comprehensive law governing all waters within its borders. The General Assembly declared that no "person or persons whatsoever, shall by wares, hedges, seines, or any other incumbrance, or means whatsoever, obstruct the natural course of passage of the fish, in the spring or proper season of the year, up or down any river in this state." Furthermore, in reference to English common law, the assembly wrote that "the same shall be deemed a common nuisance, and may be pulled down, demolished, and removed as such, by any person or persons whatsoever."[51] In 1832, some Vermont communities even proposed limiting fishing to what they called "the ordinary way," with a hook and line.[52] In 1829, Starksborough farmer-fishermen also petitioned the General Assembly to ban the use of seines and nets within the borders of the town.[53] Much like the farmer-fishermen of northern New Hampshire, these Vermont communities clearly differentiated between the destructive impacts of seines and other large nets and the comparatively benign practice of fishing with a spear or hook and line. Vermont's farmer-fishermen argued that this "ordinary way" of fishing was compatible with the overwhelming emphasis placed on maintaining the free passages that supported anadromous migrations.

In this historical moment, local consensus at the ground level among farmer-fishermen who witnessed significant fissures in migratory fish stocks gave rise to a meaningful legislative response that reflected this grassroots conservation agenda. Policymakers and fishermen designed modest restrictions to allow for a more equitable distribution of resources that simultaneously offered some hope for the possibility of sustainable river fisheries. Despite this achievement, solutions to the problem of continuous use of seines throughout the fishing season and disproportionate fishing pressure at waterfalls and at the mouths of streams and brooks remained elusive. At the end of the day, overfishing continued to be a difficult problem that placed a heavy burden on town fish wardens as the primary mechanism of enforcement.

J. W. Meader's 1869 natural history of the Merrimack River championed the industrial transformation that brought the United States onto the global stage as an economic power. With a clear commercial interest, Meader trivializes illegal fishing on the lower Merrimack River and dramatizes episodes of conflict between wardens and fishermen. He writes that fishermen regularly set their gear in illegal locations or entered the water on prohibited days, "though perhaps for the purpose of keeping the officials busy and zealous in the discharge of their duties." He continues: "Attempts were constantly made at surreptitious fishing and violation of the law, which, however, generally involved no more serious consequences than bloody noses, and the engendering of irritation and ill-feeling between the officials and fishermen." Writing in the 1860s, after river fisheries had already been compromised by industrial dams, Meader downplays efforts to prevent overfishing but does offer an anecdote that illustrates the difficulty with which wardens attempted to carry out their duties. "On one occasion, as the disciples of Walton were plying an unlawful business on Long Island (opposite the Lawrence corporation), an obnoxious and officious official from Haverhill (named Vincent), with his posse, pounced upon them, and the scene that ensued may be imagined. Donneybrook was outdone, the official and his party were repeatedly fished out of the river, after unceremonious baptisms by the faithful, and as soon as possible beat a precipitate retreat without making any arrests, but with a wholesome lesson in prudence to guide them in the future enforcement of obnoxious laws."[54]

That conflicts between fishermen and wardens often degenerated into violent encounters is perhaps understandable given the competing interests staking their claim to the fishery. Wardens represented the sole mechanism for restraining fishermen who wished to pursue their catch free from "obnoxious" state intrusion. The fish wardens reluctantly engaged in a game of cat and mouse with commercial fishing operations. As legislatures passed laws designed to prosecute malfeasance, fishermen adapted and found new ways to avoid detection.

Generally speaking, towns elected fish wardens to identify and report illegal activity to a local justice of the peace, according to the county in which the offense took place, wherein the justice issued a warrant requiring the offender's appearance in court. In Massachusetts, the justice of the peace heard testimony from both parties, at which point he rendered a verdict. In the case of a conviction, the offending fisherman either paid a substantial fine or faced imprisonment in county jail until such payment could be made. The odds did not favor the accused, as a 1790 act clearly stipulated that "the testimony of any fish warden duly chosen and sworn as aforesaid, grand juror, sheriff, deputy sheriff or constable under oath, in Court, shall be sufficient evidence to convict any person of a breach of this Act."[55] An early nineteenth-century court case in the District of Maine established a precedent that empowered states to compel citizens to abide by fisheries regulations.[56]

In 1809, Aaron Burnham, a fish warden from Scarborough, Maine, brought an offender, one Joseph Webster, to court to recoup fines related to illegal fishing on the Little River, today known as the Scarborough River, in the amount of fifteen dollars. Mr. Burnham argued common law protected his right to fish the coastal estuary, and therefore, he was not bound to respect the specific regulations that governed fishing in Cumberland County. The judge ruled unequivocally that the General Court, in this case Maine still being a province of Massachusetts, had the authority to regulate river fisheries and levy penalties for violations thereof. This case would serve as a benchmark for holding violators accountable. The precedent was clear: lawmakers implemented these regulations at the behest of the public and for the public benefit. State courts enforced the penalties that legislators found prudent to put in place.[57]

When James Cochran petitioned the Massachusetts General Court in 1811, he recognized that fish wardens stood on the front lines of the "preservation" mission. Cochran recommended a force of three wardens per town bordering the Merrimack River to identify and remove "impediments to the passage of fish," above and below Pawtucket Falls. To ensure that those who "dress themselves in disguise" would be held to account, Cochran asked that fish wardens be allowed to "seize and hold to bail all persons fishing upon an unlawful day, Who are dressed in disguise, or otherwise unknown to them, with the power to command all necessary aid for seizing and holding such offenders as aforesaid."[58] The legislature went further than Cochran might have anticipated in requiring all Merrimack valley towns to appoint six wardens to tackle a perceived uptick in illegal fishing. This statute raised penalties for fishing on unlawful days and attempted to ease the burden on fish wardens by transferring the duty for inspecting dams and other obstructions to selectmen from three towns near the obstruction. In this sense, the statute strengthened the powers of the fish wardens to catch and prosecute offenders, this being their sole remaining duty.[59] Once again, state legislatures demonstrated a commitment to side with the interests of the farmer-fishermen and the concern that dams and overfishing disrupted vital natural systems that supported New England's rural economy.

A large proportion of fisheries legislation focused on reasonable complaints that commercial monopolization displaced rural communities from this traditional resource. Unfortunately for farmer-fishermen, the history of inland fisheries management shows that as historian Richard Judd notes, "throughout the region's history, exploitation and conservation operated in tandem."[60] Conscious of those who would jettison the local consensus that formed the backbone of inland fisheries legislation in favor of immediate profit, efforts to confront destructive commercial fishing on main-stem watersheds would represent an ongoing project, undone only by the industrial transformation that put an end to the river commons once and for all. With so much emphasis on large river systems with the most abundant fisheries, fishermen managed to capture a surprising degree of local autonomy over smaller flows. With fish wardens arguably taxed beyond their limits, state authorities expressed a willingness to experiment with alternative regulatory arrangements.

Mills, Tributaries, and Town Control

During the early nineteenth century, states embraced a new management approach that placed greater control in the hands of selectmen and town officials. Riparian communities throughout the region pursued targeted actions to protect smaller flows: brooks, streams, and ponds. Where fishermen took note of the environmental impacts of milldams and destructive fishing close to home, they pressed for greater local autonomy over these smaller tributaries and, alternatively, more specific regulations geared toward the perceived needs of their freshwater ecosystems.

In 1803, the Massachusetts General Court granted the town of Haverhill authority to regulate alewife habitat in certain streams. In democratic fashion, townsfolk each year chose "by ballot a Committee of freeholders in said town, whose duty it shall be to determine and order by whom, and in what place or places; said fish may be taken in the several Streams emptying into Merrimack River within the town." To make sure the fish could freely access the upstream spawning habitat, "from tenth day of April through the month of May"—the beginning of the seasonal alewife migration—the same committee inspected nearby mills for "sluice or passage ways through or round any dams erected" and if necessary installed crude fishways consisting of "racks or wooden frames."[61] This push to decentralize inland fisheries management also challenged the wisdom of continuing to allow free and open fishing on these small alewife streams. Instead of preserving open access for all townspeople, the statute allowed the committee of freeholders "to distribute or cause to be distributed the fish that may be taken by them, or any person or persons under them, as equally as circumstances will admit," for the cost of twenty-five cents per 100 alewives, with the exception that "poor persons of said town of Haverhill . . . shall be supplied gratis with such quantities said Committee shall judge expedient.[62] In this sense, the committee of freeholders maintained equal access to fish, but not the streams themselves. Still concerned with principles of republican equality, townsfolk anticipated a future where demographic and market pressures might overwhelm the local alewife fishery if every inhabitant was guaranteed equal access to the commons. With a population of 2,730 at the beginning of the nineteenth century, perhaps this future was not far on the horizon for inhabitants of Haverhill.[63]

The Massachusetts General Court responded to the petitions of the lower Merrimack valley towns with laws that reflected a willingness to accept local solutions tailored to individual ecosystems. The most abundant of all anadromous species, alewives spawn in ponds not far from the tidewater of main-stem rivers.[64] As one of the first migratory species to leave the ocean for freshwater, sometime between March and April, this river herring entered small streams and brooks at a very opportune moment for the region's farmers. With crops still months away from harvest and supplies running low from the difficult winter season, alewives offered a nutritional bridge to the more bountiful summer season.[65] As thousands of these fish passed up small tributaries, they were taken with ease, and an emerging commercial presence within these ecosystems compelled lower Merrimack valley towns to seek new ways to exclude bad actors from their fishing grounds.

For many years, farmer-fishermen from Bradford, adjacent to Haverhill, witnessed large schools of alewives travel from the Merrimack River into Johnston's Brook, before they finally entered Johnston's and Little Ponds to spawn. This began to change when "persons from different towns employed in catching said fish" worked their nets directly above and below the entrance to Johnston's and Little Ponds. By blocking access to this spawning habitat, it was simply a matter of time until these opportunistic fishers put an end to Bradford's alewife fishery. Writing in 1802, petitioners argued that this practice, if allowed to continue, "may prove very injurious to the present and future generations."[66] Looking toward the long-term productivity of their seasonal alewife runs, they asked the General Court to relieve fishing pressure at the entrance to Johnston's Pond and, ultimately, for the power to exclude commercial actors from outside their community. The General Court first responded with a law increasing the legal fishing distance to thirty rods (165 yards) below and twenty rods (110 yards) above the stream, but then—as with the town of Haverhill—empowered an elected committee of freeholders to "oversee the taking of said fish . . . and . . . distribute the fish taken by the, or under their direction, as equally as circumstances will admit, to such Inhabitants of the said town, or other persons, as may apply for the same."[67] This petition and the resulting legislation reveal a state government willing to respect local experience by crafting additional bills with the knowledge that existing laws failed to take into account the diversity of lower Merrimack River valley ecosystems. Thus townsfolk with direct experience of their riparian environments could determine the appropriate time, location, and gear for accessing Johnston's Brook, with the rules posted in a public space for all to see.

Whether this new bill forestalled the decline of the Johnston's Pond fishery is uncertain, but these examples reveal townspeople assuming new responsibilities for the survival of a shared resource. In the case of the town of Braford, fishermen advanced a conservation vision that accounted for the needs of "future generations" and balanced the preservation of town fisheries by limiting access to the commons. Positioned as the inheritors of intergenerational knowledge, fishermen such as those from Chester, New Hampshire, leveraged claims that "for many years some of our fathers and other inhabitants of this and the adjacent town drew ample supplies of fish" to secure the right to manage local waters, in this case Cohas Brook.[68] Towns in southern New Hampshire convinced authorities to prohibit the use of seines on the Cocheco and Salmon Falls Rivers, instead permitting only the less efficient dip net.[69] The widespread acceptance of the pleas found in petitions from this period highlights the formation of a community consensus that translated into meaningful regulatory action. Rural communities advocated for targeted policies based on local environmental understandings and succeeded in relocating decision-making authority to those with direct knowledge of the issues at stake.

Conflicts over milldams as potential obstructions for anadromous migrations opened additional space for townsfolk to leverage their experience of freshwater fisheries and the threats to upstream passage. During the eighteenth and nineteenth centuries, saw- and gristmills were sited on small streams and brooks—tributaries that often led to anadromous spawning habitat. Milldams, made of some combination of stone, clay, dirt, and logs, averaged around three to four and a half meters high (ten to fifteen feet). Sluice gates controlled the flow through the milldam and directed the water, on a wooden sluiceway, or raceway, to a waterwheel that powered the mill.[70] An essential component of the agricultural economy, millers nevertheless clashed with fishermen who complained mill operations blocked access to upstream spawning habitat. Farmer-fishermen inherited a tradition built on generations of English common law in which no one had the right to obstruct the migration of freshwater species. According to such traditions, local authorities reserved the right to declare milldams, or any other obstruction, a public nuisance and order them torn down. Rural communities embraced these traditional rights as part of a larger rural consciousness that historian Gary Kulik terms "country thought." They interpreted public affairs through the lens of "power and liberty, virtue and corruption, private interests and public good." Kulik goes further, noting that country thought represented "a language of moral regeneration, inspired by classical antiquity, implicitly anticapitalist, suspicious of wealth, power, and the influence of commerce."[71] This rural ideology would prove invaluable at the turn of the century as corporate forces accumulated the political capital necessary to redefine common-law water rights, but it also informed conflicts between interconnected parties within the rural landscape.

In 1776, authorities accused millers who obstructed the passage of alewives and shad on Beaver Brook in Londonderry, New Hampshire, of "preferring private Interest, to the benefit of the Community."[72] This expression of "country thought" exposed a tension that ran beneath the surface of the rural economy. Fishermen and millers relied on the same freshwater ecosystems to secure a comfortable independence, but when millers obstructed the passage of anadromous fish, they infringed on the public fishing rights of their upstream neighbors. In the case of Beaver Brook, the miller's pursuit of "private interest" undermined the "very great advantage to the poor inhabitants & others of said towns."[73] To achieve, in the words of Charles Atkins, the "harmonizing of conflicting interests," most turn-of-the century fisheries regulations required millers to maintain a sluice opening, a fishway, or some other form of remediation to accommodate upstream passage during the spring migratory season, but this did not always satisfy interested parties. Some millers resented these prescriptions and refused to cooperate. On the other hand, fishermen regularly complained that existing fishways failed to solve the problem. Despite these tense disagreements, farmer-fishermen rarely advocated for the complete removal of offending dams. Instead, they asked for meaningful alterations that would allow spring fish runs to coexist with the water needs of local mills.

In 1788, town selectmen from Winchester, New Hampshire, pressed for new measures to regulate milldams on the Ashuelot River, a tributary of the Connecticut watershed located in southwestern portion of the state. As the largest tributary of the Connecticut River within the borders of New Hampshire, the Ashuelot supplied households with "a large number of Salmon and Shad." According to the petitioners, the fishery declined "by reason of the Dams across said river in Hinsdale, Winchester, and Swanzey." They pushed for a new "act that shall open a free course for said fish," so "that the inhabitants may receive the benefit of the same."[74] The New Hampshire House of Representatives referred the petition to a committee, which asked the mill owner to testify before the body. In this sense, the state legislature placed the burden on mill owners to demonstrate why they should operate free from regulation.[75] Ultimately, New Hampshire granted the petition and passed a law that required all dams on the Connecticut and Ashuelot Rivers to open a passageway between May and July each year. In addition, it tasked the three towns with the formation of a committee to inspect milldams in order to ensure compliance with the law.[76] This seems to have been a relatively simple conclusion to a seemingly intractable disagreement, yet farmer-fishermen, wardens, and inspection committees often took issue with the notion that fishways represented the universal solution. In other cases, farmer-fishermen applied a compelling new natural philosophy in arguing their case.

In July 1790, petitioners from Gilmanton and Meredith, near Lake Winnipesaukee in New Hampshire, declared that no milldam on the Merrimack River should extend "more than one half the way across the river" and that the other half should remain "free as nature made it."[77] The selectmen and justices of the peace from both towns engaged in a bitter two-year dispute with Major Ladd, a particularly obstinate mill owner, and the episode convinced petitioners of the utility of this new approach. In May 1789, town officials visited Major Ladd's dam, situated on the Merrimack River between the two towns, and unanimously declared "we are all of the opinion that the dam obstructed the fish." As a result, they "ordered Ladd to take part of it down, but he paid no attention to our order." The officials went back and forth with Ladd for the next year, as Ladd appealed court decisions and delayed judgments to comply with lawful orders. In the meantime, informants accused him of surreptitiously attaching flashboards to raise the height of his dam, in which case "sluice ways will never preserve the fish if they were allowed to keep a dam two or three feet high . . . they would have it so contrived as to put up flash boards at any time so as totally obstruct the passage of the fish." Exhausted by the amount of time and resources devoted to addressing this single obstruction, town officials settled on the far more radical intervention that kept open channels "free as nature made it."[78] A partial return to the natural ecology of the river would provide anadromous species far more space to maneuver upstream than a small opening through Ladd's dam. Fishermen and town officials from the upper Merrimack River valley leveraged this environmental ethic to protect both the river's flow and the interconnected public rights to the river commons.

Major Ladd, for his part, continued to resist any perceived interference with his livelihood. He complained that the selectmen and justices could not be trusted to determine what might constitute a "sufficient sluice way." He further asked that "a suitable committee who are well acquainted with the river Merrimack" take up the matter and leave him free to operate his mill as he saw fit. It is unclear how the conflict was ultimately mediated, but this episode speaks to fissures within the rural economy that separated the "private interests" of mill owners from community fishing rights grounded in law and custom. Town officials, likely fishermen themselves, held to a vision of river connectivity that they believed established a greater service to riparian communities as a whole. Unfortunately, many rural inhabitants did not subscribe to this way of thinking. As a result, it does seem that, much like the dams themselves, at least some of these disputes were insurmountable.[79]

In other examples, states took a far more proactive approach in sorting through competing demands on freshwater ecosystems. In Chelmsford, Massachusetts, Stoney Brook travels nearly twenty-two miles northeast before emptying into the lower Merrimack River. In June 1804, fish wardens from Chelmsford and Westford, performing the duties of their office, inspected William Adams's milldam and testified that his sluiceway failed to accommodate the passage of alewives over his dam. Adams rejected this determination and petitioned the General Court to continue the operation of his two gristmills and two sawmills. He emphasized his position in the local economy, writing that his "mills are importantly useful to your petitioners who cannot be so well accommodated by any other mills . . . being situated on the banks of the Merrimack the lumber that is sawed there can be . . . transported to the Capital by water and also . . . the grist mills are greatly used by people carrying grain to market."[80]

Dozens of signatures appear on this petition, many of which were certainly those of commercial producers who relied on these mills to prepare timber and agricultural products for sale in Boston. Farmer-fishermen would also require Adams's services to grind their cereal grains into flour or to acquire lumber for annual farm improvements. Adams understood his strong position in both the market and rural economies. He pointed out that with no other facilities operating within a reasonable traveling distance, the fish wardens would leave the community in a precarious state with all four mills temporarily suspended during the construction of a new fishway. He wanted to continue to operate, "provided he shall keep the sluice way in good and sufficient repair and open for the passage of fish as long as is now required by law."[81] The fish wardens continued to assert that Adams's milldam was "altogether insufficient and fish will never ascend, wherefore it being highly advantageous to the Said towns and other towns that the shad, alewives, and other fish should ascend the said brook, which were it not for said mill and dams they would freely do."[82]

To resolve this impasse, the General Court formed a special committee to once again inspect Adams's milldam and issue a final judgment. Much to the surprise of the fish wardens, the special committee sided with Adams.[83] They settled on a temporary solution that permitted Adams to work his grist- and sawmills for a term of three years without costly improvements to the existing fishway. The committee also communicated a willingness to revisit the matter should the wardens' misgivings prove to be true.[84]

In the case of Stoney Brook and Adams's dam, the General Court had the difficult task of reconciling a productive freshwater habitat with the built environment. Gary Kulik argues that fair-handed mediators could intercede to resolve these disputes since rural mill owners did not represent a separate industrial interest.[85] Early nineteenth-century mills served both local farmers and the broader market economy. In the end, it was in Adams's own interest to avoid alienating local farmers, many of whom annually fished for alewives in Stoney Brook, since their patronage also kept his business in operation. Conversely, nearby farm households depended on the services of Adams's mills. They both represented discrete but interdependent components of the rural landscape.

From time to time, dam obstructions forced townsfolk to take matters into their own hands. At the turn of the nineteenth century, residents of Derryfield, New Hampshire, present-day Manchester, expressed alarm at the decline of fish in nearby Cohas Brook. The petitioners believed that the disappearance of alewives resulted from the numerous milldams that dotted this small Merrimack River tributary. After first securing permission from the New Hampshire General Court, the fishermen described that "they with cheerfulness and alacrity caused the aforesaid obstructions to be removed."[86] English common law empowered colonial fishermen to tear down dams that blocked anadromous fish run, so this type of property destruction, though rare, was not entirely without precedent.[87] Townspeople rarely took part in violence without the sanction of legal authorities, but historian Theodore Steinberg documents a dramatic case of extralegal retaliation against large dams in northern New Hampshire. In 1859, the Boston Associates, the investors behind the Waltham-Lowell system responsible in good part for the growth of industrial manufacturing in the United States, built a series of storage dams within the upper Merrimack watershed to create an even and consistent flow of water power for textile mills further south in Lowell and Lawrence. One such dam at the headwaters of the Merrimack River, near the southern end of Lake Winnipesaukee, drew the ire of locals that resented the presence of these elite industrial interests from outside their community, especially as their manipulation of the local environment resulted in flooded meadows and tillage. When residents attempted to tear down the dam, the situation degenerated into blows between landowners and private security forces. This dramatic event foreshadowed a growing socioeconomic divide between agricultural and industrial interests.[88] Disputes over water rights would continue to escalate as industrialization displaced traditional relationships with riparian environments.

Conclusion

In working to confront destructive overfishing and clear the way for anadromous migration, farmer-fishermen articulated a nascent conservation ethic that gave rise to an ongoing legislative program to preserve inland fisheries. Legislative documents reveal that farmer-fishermen and local authorities understood the fragile conditions that supported seasonal migrations from marine to freshwater habitat and allowed anadromous species to complete their life cycle. Witnessing the impacts of commercial overexploitation and obstructions from small dams, farmer-fishermen, wardens, town selectmen, and justices of the peace collaborated with state lawmakers to sustainably manage river systems as a common-pool resource. Toward this end, farmer-fishermen pushed for use limits that stipulated the accept-able time, location, and manner one could access the river commons. These prescriptions originated not only from local knowledge of freshwater ecosystems but also from a moral tradition that stressed equal access to resources that supported economic independence in the countryside. Petitions and legislation characterized access to river fishing as either a great service, a public benefit, or a public good.[89] The rhetoric of community rights lent moral weight to decades of petitions sent to state authorities.

Between 1783 and 1820, state legislatures responded to the peti-tions of rural communities with measures to address the decline of fisheries on large, main-stem river systems and smaller brooks, streams, ponds, and other freshwater spawning habitat. To con-front an expanding commercial presence across New England's river systems, farmer-fishermen secured protections for vulnerable habitat that increasingly attracted overwhelming amounts of fishing pressure. Above and below large waterfalls where salmon and shad schooled in large numbers, as well as near the mouths of smaller brooks and streams that led to alewife habitat, they pursued restric-tions on the size and location of seines and large nets. In one of the more compelling examples of interstate cooperation, during the late eighteenth century, fishermen and state legislators reached a tentative consensus that established a closed season for large nets on the Con-necticut River. Despite these achievements, persistent complaints of destructive fishing forced stakeholders to constantly revisit the diffi-cult question of how best to exclude bad actors. As James Cochran identified in 1811, this almost impossible task mostly fell to town fish wardens, who were collectively responsible for policing hundreds of miles of freshwater habitat while offenders disguised, deceived, and overfished. Stakeholders made inroads in mitigating overfishing, but the commons dilemma would not be so easily resolved.

From the colonial period to the early nineteenth century, state legislators proved a strong ally in the defense of sustainable river fisheries. They expressed common cause with riparian communities with legislation derived from locals who understood the intricacies of their rivers and streams. Presented with compelling arguments for greater local control, lawmakers accommodated town officials who wished to regulate smaller flows. When necessary, legislative committees stepped in to mediate conflicts between fishermen and mill operators, though perhaps not to the satisfaction of all interested parties. Across several decades, legislators showed a willingness to amend, renew, and reform the regulatory scaffolding based on the testimony of stakeholders. The evolving nature of this conservation agenda points to a process of experimentation in which farmer-fishermen and state legislatures worked together to identify the correct balance of regulation and enforcement to prevent the monopolization by commercial fishers, or part-time opportunists, who sought to maximize their catch at the expense of future returns. It is likely that this era of regulatory experimentation would have continued, due to the public pressure put on state legislatures, had the arrival of industrial dams been delayed or the dams themselves made passable. On large main-stem watersheds like the Connecticut and the Merrimack Rivers, industrial dams interrupted a sustained, decades-long effort to reign in commercial overexploitation and preserve subsistence fishing customs. While farmer-fishermen continued to guard community rights to inland fisheries, by the early nineteenth century they faced existential threats as industrial interests redefined traditional water rights to power a new manufacturing economy.

CHAPTER FIVE

"FROM TIME IMMEMORIAL"

Closing the Commons

In August 1839, Henry David Thoreau and his brother, John, launched their boat into the Concord River and followed the gentle current to the confluence of the Merrimack River. Paddling upstream against the current, as he had done all his life, Thoreau took note of changes in the land and water. The farmhouses, barns, and meadows that formed Thoreau's memories of the Concord River Valley and the surrounding Massachusetts countryside slowly gave way to canals, large dams, steamboats, and bustling factory towns. These fixtures of the new commercial and industrial age would soon overshadow an Indigenous and colonial environmental heritage, as Thoreau describes: "Pawtucket and Wamesit, where the Indian resorted in the fishing season, are now Lowell, the city of spindles, and Manchester of America, which sends its cotton cloth round the globe."[1] As politicians, judges, and a large swath of the general public championed this new global industrial power, Thoreau warned that this period of dynamic socioeconomic transformation was not without environmental and spiritual costs. New England's small agricultural communities understood this better than anyone. Farmer-fishermen cautioned that a new world dominated by market forces would displace a traditional subsistence economy, predicated on diffuse, seasonal labor—and partially supported by seasonal fish migrations. While the small minority of citizens that subscribed to the natural philosophy of Thoreau and Emerson sought to recapture some small measure of what had been lost, forty years earlier farmer-fishermen vigorously defended what they feared was a way of life under attack.

On the eve of the Industrial Revolution, few could have guessed the radical ways in which the foundations of early American society would be inexorably altered. In New England, large-scale textile manufacturing, based on the wildly successful Waltham-Lowell system, harnessed the raw energy of the region's most powerful rivers. Although historians have focused on the far-reaching consequences for the nation as a whole, they have largely overlooked those who experienced these changes at the grassroots level. Small agricultural communities maintained traditional rights to rivers informed by the norms and values of the rural economy, and it was precisely these communities that rose to confront large-scale dam proposals. The appearance of these new industrial projects throughout the region offers a compelling story of tension, conflict, and adaptation in which rural communities formed coalitions to protect traditional rights to the river commons. They resisted the transition to an industrial economy, and when controversial proposals to construct large dams threatened to drown fishing grounds beneath massive reservoirs, rural communities called on state governments to intervene. In the face of wholesale ecological transformations, fishers, millers, and riparian landowners came together in defense of public rights.

Before the arrival of industrial dams, state authorities and private investors manipulated the natural course of large river systems to improve navigation and expand commercial enterprise. Policymakers and private entrepreneurs engineered locks and canals to circumvent falls, rapids, and other natural obstacles, thus enabling freight and passenger boats to reach communities previously cut off from regional commerce. The Proprietors of the Locks and Canals on the Connecticut River funded the construction of the South Hadley Canal, the first navigable canal in the United States, which opened in western Massachusetts in 1795. Subsequent canal projects at Millers and Turners Falls extended the navigable reach of the Connecticut River from Long Island Sound to Vermont. By 1810, commercial vessels could access 250 miles of the watershed.[2]

FIGURE 5. *Canals of the Nineteenth Century*, in William F. Robinson, *Abandoned New England: Its Hidden Ruins and Where to Find Them* (Boston: New York Graphic Society, 1976), 21.

These innovations brought the Merrimack River and Concord River valleys within the commercial orbit of Boston. Opened in 1802, the Middlesex Canal represented a breakthrough commercial achievement and foreshadowed dam projects that would continue to stretch the limits of nineteenth-century technology. Beginning

in Chelmsford and then following the Concord River south to Billerica, this three-and-a-half-foot-deep ditch, complete with a series
of twenty locks, allowed boats to travel twenty-eight miles south to
Boston Harbor.[3] To supply the canal with water, project managers
rebuilt and raised an old dam in Billerica. Unfortunately, the backwater behind the dam, according to Thoreau, flooded thousands
of acres of river meadows, the backbone of the local agricultural
regime. Historian Brian Donahue writes that Concord River valley
farmers "claimed that because the dam slowed drainage, the river
flooded more frequently and stayed flooded longer following summer rains, spoiling the hay crop."[4] Before the canal project, these
meadows fed cattle, which in turn fertilized planting fields—for a
time, a sustainable closed nutrient system. In the face of this environmental transformation, Thoreau continued to represent the traditional intersection of farming and fishing in the Concord River
Valley: "At length it would seem that the interests, not of the fishes
only, but of the men of Wayland, of Sudbury, of Concord, demand
the leveling of that dam." Despite Thoreau's protest, many farmers
embraced the new commercial opportunities brought about by the
Middlesex Canal, gave up on cereal grains, specialized their production for growing factory towns, and increasingly relied on regions
outside the Northeast to furnish needed farm supplies.[5] Thoreau's
world was changing, along with that of the Concord Valley farmers
and fishermen, but he still clung to romantic memories of "the same
fisher in my earliest youth, still as near the river as he could get,
with uncertain undulating step, after so many things had gone down
stream, swinging a scythe in the meadow, his bottle like a serpent
hid in the grass; himself yet not cut down by the Great Mower."[6]

Canals opened the floodgates to a series of technological innovations that further compromised river ecology and the subsistence
values of New England's rural economy. Between 1814 and 1824,
the introduction of the power loom, the double speeder and filing
frame, and the self-acting loom temple ushered in a new era of textile manufacturing defined by increased efficiency, higher productivity, and, above all else, record profits. Combined with a large-scale,

integrated production system, later termed the "Waltham-Lowell system," and pioneered by new, highly capitalized corporate firms such as the Boston Manufacturing Company, New England would lead the way in American textile production for decades to come. Between 1800 and 1829, Massachusetts issued 148 corporate charters to manufacturers.[7] Smaller but more numerous firms, located on streams across southern New England, particularly in Rhode Island, also contributed to this new enterprising spirit.[8]

Corporate interests launched this manufacturing revolution in part by leveraging centuries-old colonial legislation to redefine common-law water rights. Beginning in 1714, the Massachusetts General Court passed a series of acts designed to encourage mill construction and grow the agricultural economy. Under the "Mill Acts," riparian landowners who found their land flooded as a result of small dam construction could collect a modest compensation, but ultimately their rights as property owners were sacrificed to a common interest in economic growth.[9] The laws included special protections for fish that required open passages for seasonal fish migrations, but this would prove difficult to achieve with industrial dams, given their size. Firms modeled on the Waltham-Lowell system required massive amounts of water power that necessitated large industrial dams on main-stem rivers like the Connecticut, Merrimack, Kennebec, and Penobscot. The Essex Dam in Lawrence, the largest of any in the United States or Europe at the time it was finished in 1848, represented an impassable obstacle with its thirty-two feet of cement stretching upward from the river bottom across the entire width of the Merrimack.[10] When industrial manufacturing began to take off in the 1820s, corporate interests benefited from judicial determinations that extended the protections of the Mill Acts to industrial manufacturing.

Lemuel Shaw's appointment as chief justice of the Massachusetts Supreme Judicial Court in 1830 precipitated a stark reinterpretation of common-law water rights. Legislators and judges had long prioritized traditional fishing rights given the overwhelming consensus associated with the preservation of the river commons, but over the course of his thirty years on the state supreme court, Shaw consistently decided cases in favor of manufacturing interests. One of the most consequential jurists in American history, Shaw's judicial philosophy elevated economic growth above traditional property and water rights.[11] Legal scholar Morton Horwitz describes this moment as one of transition in which "the anti-development doctrine of the common law first clashed with the spirit of economic improvement." Common-law tradition had protected anadromous fish runs for hundreds of years, but as Gary Kulik explains, "claims of unhindered economic development came to outweigh both the customary precepts of property law . . . and the claims of public rights."[12] With new corporate entities empowered by the courts and lawmakers to dam rivers and grow the industrial economy, conflicts over water rights degenerated into a zero-sum game.

In the case of the Merrimack River, the stage was set for conflict. The farmer-fishermen of rural Massachusetts refused to cede their rights to harvest the seasonal migratory fish runs. They fought tirelessly against efforts to monopolize the river, winning some important victories along the way, but as industrial manufacturing accumulated wealth and political influence, commercial operations fished more intensively to meet rising demands for offshore bait and domestic consumption. Given these twin pressures, the decline of the fishery was all but ensured. The Boston Associates' large dams at Lowell and Lawrence, built in 1830 and 1849, respectively, catalyzed the expansion of industrial manufacturing in the United States, but they also destroyed the river fishery.[13] However, the way agricultural communities responded to these new pressures foreshadowed renewed commitments to restoring migratory fish in the 1870s and again in the 1970s.

Massachusetts: Defending the Commons

Beginning at the confluence of the Pemigewasset and Winnipesaukee Rivers in Franklin, New Hampshire, the Merrimack flows for 127 miles before reaching the Atlantic Ocean in Newburyport, Massachusetts, draining an area of 5,014 square miles. After flowing south through New Hampshire, the river takes a sharp turn toward the northeast,

where its final forty-nine miles are often referred to as the lower Merrimack watershed.[14] This portion of the river has often drawn the attention of historians because the lower Merrimack River offered fertile
grounds for the expansion of urban, commercial, and industrial growth
associated with the Market Revolution. Prior to the industrial success
of the Boston Associates, two separate business interests attempted to
dam the Merrimack at Pawtucket Falls. Local agricultural communities, aware of the potential threats to migratory fish runs, vigorously
resisted this transition to a manufacturing economy.

The controversy began in the late eighteenth century when a group
of prominent merchants from Newburyport incorporated as the Proprietors of Locks and Canals on the Merrimack River. They funded
a canal around Pawtucket Falls—an unnavigable thirty-two-foot
drop—which allowed freight vessels from above the falls to sail down
to the mouth of the river at Newburyport and the Atlantic Ocean.
Within a few years, the nearby Middlesex Canal, which transported
goods from the Merrimack valley directly to Boston, essentially put
the proprietors out of business. This forced them to seek other means
to profit from the river. Dudley A. Tyng, a lawyer from Newburyport
and an original shareholder of the Proprietors of Locks and Canals,
petitioned the Massachusetts General Court for permission to construct a dam across the falls.[15] This represented the first large-scale
effort to control one of the most powerful points of the Merrimack
River, as well as the most vulnerable in terms of migratory fish runs.
As a result, several towns from the lower Merrimack valley organized
a successful grassroots opposition to Tyng's efforts.

In June 1801, the inhabitants of Dracut gathered in a town meeting
to discuss this proposal. In a remonstrance protesting this project, they
denounced Tyng's dam as a disaster for migratory fish and the communities that relied on this resource. They called on legislators, as they had
in the past, to account for the potential threats to community rights and
the public good. "The erection of a dam at Pawtucket falls, in the manner proposed by petitioners . . . will in the opinion of the town totally
destroy the fish in said river, and deprive the people of the important
privilege which they for a long time, even from time immemorial, have

enjoyed without molestation, of taking near their doors, the most delicate food and much of the real necessities of life, and no other purpose
can be offered, than a gratification of the avaricious feelings of a few
individuals who must be unacquainted with the real effect of this measure, or regardless of the public good."[16] In evoking the phrase "from
time immemorial," these farmer-fishermen embraced an environmental heritage, a relationship with the Merrimack River, well established
in common law and colonial history. Agricultural communities along
the Merrimack River understood what was at stake. This declaration
drew a clear line: on one side, the many farmers who relied on migratory fish to sustain a comfortable subsistence, "the necessities of life";
on the other, an "avaricious few" manufacturers who sought to transform the landscape, monopolize the water, and in the process deprive
locals of an essential natural resource.

FIGURE 6. *Pawtucket Falls*, by R. E. Westcott, circa 1896, in *Picturesque Lowell* (Lowell, MA: T. H. Lawler, Bookseller and Stationer, 1896). Prints and Photographs Division, Library of Congress, Washington, D.C., LC-DIG-ppmsca-17295.

The demographic blueprint of the lower Merrimack valley ranged from thriving coastal cities, such as Salem and Newburyport, to small agricultural towns further inland. These latter communities, such as Dracut, Chelmsford, Westford, and Bradford, all possessed populations under 1,400 people. Unlike coastal Salem and Newburyport, with populations of 9,457 and 5,946, respectively, most towns of the lower Merrimack valley inhabited a mixed rural landscape in which a myriad of activities offered a path to economic independence and prosperity.[17] George Brown Goode's 1880 report, *The Fisheries and Fishery Industries of the United States*, noted that several fishermen could still remember the Merrimack River teeming with alewives, shad, and eels.[18] When outside interests threatened to undermine that component of the rural economy, towns throughout the river valley rose up in protest.

The town of Bradford continued the campaign of petitions, pointing to the certain destruction that would result from damming the entirety of Pawtucket Falls. Nearly thirty miles downriver from the falls, inhabitants of Bradford understood that should this dam obstruct migratory fish from ascending the falls in search of spawning grounds, they could not rely on sufficient returns the following season and the fishery would soon collapse. With this in mind, they wrote their petition, "feeling the importance and very great advantage that the present generation, and much more that of posterity, may derive by the passages for the fish that usually go up Merrimack River being kept clear." Similar to their neighbors upriver in Dracut, petitioners from Bradford denounced the "selfish motives, to obstruct those passages for private emoluments, and at present among the most destructive of those plans."[19] With this language, Bradford farmer-fishermen interpreted Tyng's proposal for a dam as a struggle to secure communal rights from corporate impositions not only for themselves but also for future generations. These types of zero-sum conflicts between community and corporate interests would continue to inflame socioeconomic divisions into the industrial era.

With the understanding that the General Court had long accepted the advice of local townspeople and fish wardens, the Bradford petition put forward an argument based on trusted accounts: "We find from experience that the wise and wholesome laws for regulating the fishery on Merrimack River in the present partial execution . . . has in some measure retrieved the almost exhausted run of fish in said river."[20] Efforts to protect the river fishery, though partially successful, would be entirely negated by the dam proposal. Bradford townspeople claimed their experience of the local environment gave them the authority to speak on such consequential matters, not to mention the importance of migratory fish to the livelihood of rural Merrimack valley communities.

The question of the dam construction at Pawtucket Falls was then put to the citizens of Andover at a town meeting in the same month. They voted unanimously to send a representative to the General Court so that their opposition to the proposed construction might be made in person.[21] Around the same time, the town of Chelmsford held a meeting and "after taking the matter under consideration the town unanimously voted to remonstrate against said petition." They noted the very real possibility "that it will be a very great damage to the fishery in that river, if it does not wholly destroy it." Residents of Chelmsford again predicted the certain destruction of the fishery upstream from Pawtucket Falls as a result of this project but took a more expansive view of the potential environmental harms. Referring to alluvial flats where farmers harvested meadow hay, gathered wild foods, and hunted game, petitioners believed Tyng's dam was likely to flood these "intervals for considerable distance above said dam."[22] Though the survival of the fishery remained at the center of the argument, farmer-fishermen strategically pointed to the diverse, but interconnected, natural resources within the lower Merrimack River ecosystem that supported agricultural communities.

Perhaps most interesting, the document also refers to the Stoney Brook mills, so prominent in the early fisheries debate on the lower Merrimack. Residents of Chelmsford warned that this dam would restrict the river's flow and prevent the Stoney Brook mills from operating at full capacity.[23] This tacit coalition among farmer-fishermen and small mill operators underscores the looming threat of industrialization for agricultural communities. While these groups may have squabbled in the past and would continue to do so in the future, they ultimately recognized the webs of dependence that held the rural economy together. Proto-industrial interests, in their quest to redefine the most productive application of water resources, represented the true threat to the community and the public good.

Dracut townsfolk believed that the project would not only destroy the migratory fish populations of the Merrimack but also deprive rural communities of an important privilege that they had enjoyed "from time immemorial." Going even further, they maintained that access to the Merrimack River fishery also represented a necessary aspect of the economic landscape, "the most delicate food, and much of the real necessaries of life." Tyng and his associates, they argued, must have known the devastating consequences of their dam proposal, especially for the thousands that relied on fishing as an integral part of their livelihood. Fishermen worried that the "gratification of the avaricious feelings of a few individuals" might supersede the "public good."[24] Again, the tension between community and individual rights stood at the forefront of this resource conflict. By arguing that access to the Merrimack fishery was a matter of common law and custom, these towns emphasized the firm precedent established by more than one hundred years of settlement in the valley. In the end, a combination of rural interests successfully defeated efforts to construct a dam across Pawtucket Falls, creating a logjam in the legislature that frustrated Tyng's efforts to secure the necessary permissions. These towns successfully convinced the General Court of their authority to speak on the subject of river fisheries and the consequences of human interference. Nevertheless, two years later, this issue would again capture the attention of Merrimack valley farmer-fishermen.

In May 1812, John Ford, a large landholder, mill owner, and entrepreneur operating under the auspices of the proprietors of both the Pawtucket and the Middlesex Canals, also sought permission from the General Court to construct a dam across Pawtucket Falls. Several towns spoke out in outrage and again identified a thinly veiled

effort to capture a public resource for private interests. As petitioners from Chelmsford understood, this proposal was calculated to
"accommodate two rich and wealthy corporations at the expense
and injury of many honest and industrious citizens."[25] Such a bold
defense of preindustrial class values appealed to suspicions that while
a new manufacturing economy promised wealth and prosperity, it
also led to pronounced socioeconomic divisions. Disconnected from
the resources that sustained the "honest and industrious citizens" of
the lower Merrimack valley, rural communities demonstrated a keen
sense of the shifting currents that might potentially favor a new manufacturing class. Like every petition from the previous twenty years,
fishermen reiterated that the salmon, shad, and alewives "are of great
value to the town and adjacent county, and the business of catching
them is a great advantage to the poorer class of people and should
the dam prayed for be erected . . . in a very few years no fish will be
caught in that river."[26] The petitioners understood the precarious
nature of inland fish migrations and identified precise locations that
allowed fish an opportunity to surmount the obstacle. They anticipated the dam would block an area "very narrow between ledges, at
the head of the falls and the water falls never then with great violence
and velocity. If the dam is erected across then it will be impossible
for any fish to pass the falls." Fishermen claimed special knowledge
of this local environment and played up their credibility to overcome corporate initiatives from outside the community. Not only
would the migratory fish be compromised, inhabitants of Chelmsford argued, but since 1698 "a gristmill has been kept in good repair
on that privilege, which will entirely be destroyed by back water if
a dam be erected across Pawtucket Falls."[27] The gristmill had operated "for the use of said town" as far back as the seventeenth century.
In this sense, the proposal would have had a profoundly negative
impact on two essential facets of New England's rural economy. The
town of Chelmsford was not alone in this critical assessment.

Bradford residents yet again raised their collective voice in challenging this dam proposal. They called on the legislature to consider the spiritual forces that set these natural processes in motion.

Decades later, Thoreau and the Transcendentalists would promote a strikingly similar natural philosophy. The petitioners wrote, "It is well known that by function which the great architect of creation, have in his wisdom seen fit to make in that place, is such as to render it extremely difficult for the shad fish, and even the beautiful salmon with his great strength and activity, to ascend that rapid cataract." The "great architect of creation" opened these small channels for shad and salmon to overcome the falls, and therefore, it was incumbent on these fishermen to act as good stewards of the land and water. Such a departure from the natural order, that would serve to disrupt these divinely inspired migrations, would then represent an affront not only to the fish and fishermen but also to the Creator as well. This line of reasoning seems to proceed from the knowledge that a good number of Jesus's disciples were, in fact, fishermen. The Bradford petitioners then brought the defense back to the material well-being of lower Merrimack communities, as they described that "the inhabitants living on the margin of said river annually procure large quantities of those fish for the sustenance of their families."[28]

While the argument for protecting inland fisheries remained uppermost in the minds of these farmer-fishermen, William Adams, the mill operator on Stoney Brook, threw his support behind the effort to defeat this obstruction. Adams appealed on the basis of history in that "for more than one hundred years there have been mills in operation at the same place where his now stands," where Stoney Brook empties into the Merrimack River. Adams insisted that Ford's dam would compromise his grist- and sawmills, "for the want of water sufficient to work them." He echoed the farmer-fishermen's defense of community rights, "to grant a privilege to an individual which would at the same time deprive others of a privilege far more valuable." He concluded with the simple assertion that "the public good does not require " the large dam.[29] Back in 1804, William Adams had squabbled with fishermen and wardens in relation to his own small milldam, an episode highlighted in the previous chapter, but the appearance of external commercial interests united millers and fishers in a tacit coalition against this industrial takeover of the river. Lower Merrimack valley communities realized the environmental impacts of the proposed Pawtucket dam would not be limited to a single constituency; rather, a variety of rural interests shared a stake in keeping the main-stem channel free from industrial obstructions.

The nearby counties of Rockingham and Hillsborough, New Hampshire, also drafted remonstrances objecting to Ford's dam proposal.[30] They calculated that the environmental impacts would flow upriver to New Hampshire farmer-fishermen and lead to the destruction of the river fishery in the upper reaches of the watershed. Historic fishing sites such as Amoskeag Falls would be left as monuments to a forgotten past. This defense by small mill operators, riparian landowners, and fishermen from both Massachusetts and New Hampshire speaks to an integrated preindustrial world rather than a simple plea to preserve the fisheries. In February 1813, John Ford withdrew his proposal, but two attempts at industrializing the waters of the Merrimack River in the span of twelve years indicated the unsettled nature of the issue. A coalition of rural interests won the first two rounds, but the influential investor class in Massachusetts worked diligently to bring politicians to their way of thinking. A decade later, Francis Cabot Lowell, Kirk Boott, and Nathan Appleton would eventually succeed where others had failed.

While it is important to recognize the efforts of farmer-fishermen to resist the closing of the river commons, fishermen were not always so attuned to the fragile ecological conditions that sustained productive inland fisheries. Deep fissures in Maine's fishing communities undermined the development of a consistent legislative program for preservation. Maine fishermen fought with one another and with state regulators over issues of equipment, fishing location, and the inclination toward commercialization as opposed to community-based country fishing, but the one thing that most fishermen could agree on was that fish stocks were in decline. In the case of the Penobscot and Kennebec Rivers, the absence of a broad-based coalition to preserve inland fish stocks brought the resource to the edge of collapse but also paved the way for industrial dams.

"Our equal right to fish"

Shortly after Maine achieved statehood in 1820 as part of the Missouri Compromise, policymakers found themselves inundated with distressing reports about the poor condition of the state's inland fisheries. Fishermen scrambled to assign blame to their competitors. Weir fishermen represented a powerful commercial interest within the tidewater of Maine's largest watersheds, and they refused to entertain the possibility that their operations contributed to overexploitation. Communities further upriver disagreed. In 1820, Elijah Johnston and other fishermen described the Penobscot River fishery as a "source of publick benefit," asking the Maine Legislature to restrict the number and location of weirs that they believed had "destroyed the said fish at all seasons of the year, both old and young fish so that they have become almost extinct."[31] Mr. Johnston was not alone in that opinion. The next year, the Indigenous peoples of the Penobscot Nation wrote that "in the days of our forefathers the great plenty of fish which yearly came into the waters of our Penobscot River was one of the greatest sources by which they obtained their living and has so continued." This all changed "when our white brethren came amongst us, they settled on our lands and near the tide waters of our River . . . have every year built so many wears that that they have caught and killed so many of the fish that there is hardly any comes up the River where we live." This kind of overfishing not only affected the Penobscot's material subsistence but also the foundations of tribal culture, disconnecting their community from a resource they "always considered as sent us by the great god who provides for all his children." The spiritual traditions of the Penobscot people stood in stark contrast to the commercial ethic of weir fishermen in the lower river. Indigenous environmental knowledge, embedded within animistic spiritual traditions, encouraged reciprocity and respect between the human and the nonhuman worlds. Toward that end, the Penobscot called for sustainable practices aligned with their traditional "use of very small nets and spears" and legislation not only to prevent the construction of additional weirs but also to prohibit "white people and Indians from catching fish more than two days a week in the season of salmon, shad, and alewives at least for five years." Penobscot commitments to sustainable use extended to both their "white brethren" and their own fishing activity until a time when "fish will then be plenty again."[32] The Penobscot's petition emerged from a distinct cultural context, but the guiding principle was the same for both Native American and Euro-American inhabitants of the Penobscot River. With the banks of the lower river crowded with weirs, fishermen in the estuary trapped large schools of fish to the detriment of communities further upriver. Left unchecked, these profligate harvests would soon put an end to the fishery. To secure equal access to this resource, commercial harvesters, whose interaction with the river commons damaged the rights of stakeholders further upriver—particularly the Penobscot Nation— would have to be restrained.

On the lower Penobscot River, commercial fishermen employed two types of fixed equipment to maximize the catch. The "floored weir," most common on the Penobscot, featured a brush fence extending straight from the river bank into the channel known as the "leader." The leader guided the fish toward three increasingly narrow, heart-shaped "pounds" made of brush and netting. Fishermen attached the outer two pounds to the muddy river bottom and built wooden platforms, or floors, that rose just above the low watermark. As the tide went out, the netting trapped fish on the exposed wooden platform. Pound nets, used commonly on the west side of the bay below Belfast, closely resembled weirs but sat atop wooden floats and used net with large mesh to snare fish as they attempted to swim through. Maine fish commissioner Charles Atkins estimated that each weir required an initial investment in materials and labor of eighty to one hundred dollars, with annual upkeep ranging from fifty to eighty dollars.[33] There is no doubt that too much fishing pressure within the Penobscot's tidewater, particularly weir fishermen, contributed to significant declines in salmon populations. Nevertheless, strong financial incentives and substantial capital investments motivated this type of intensive fishing. With such significant outlays, weir fishers pursued commercial opportunities that far outstripped the scope and scale of community fishing further upriver.

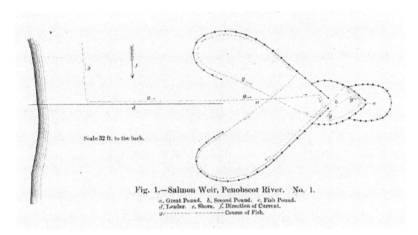

Fig. 1.—Salmon Weir, Penobscot River. No. 1.
a. Great Pound. *b.* Second Pound. *c.* Fish Pound.
d. Leader. *e.* Shore. *f.* Direction of Current.
g. - - - - - - - - - - - - - - Course of Fish.

FIGURE 7. *Salmon Weir, Penobscot River. No. 1,* United States Commission of Fish and Fisheries, *Report of the Commissioner, 1872–1873,* Part II, Plate XII, (Washington, D.C.: Government Printing Office, 1874).

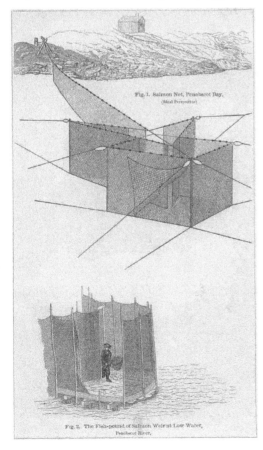

Fig. 1. Salmon Net, Penobscot Bay.
(Ideal Perspective)

Fig. 2. The Fish-pound of Salmon Weir at Low Water,
Penobscot River.

FIGURE 8. Fig. 1, *Salmon Net, Penobscot Bay,* and Fig. 2, *The Fish-pound of Salmon Weir at Low Water, Penobscot River,* United States Commission of Fish and Fisheries, *Report of the Commissioner, 1872–1873,* Part II, Plate XI (Washington, D.C.: Government Printing Office, 1874).

For the first half of the nineteenth century, ships from southern New England regularly sailed up to the Kennebec and Penobscot to purchase barrels of salted or smoked salmon for urban seaport markets. By 1850, with more sophisticated preservation methods and rail networks connecting Rockland, in the outer bay, to Portland and points south, fish dealers packed fresh salmon on ice, which quickly caught on with retailers and consumers in Boston.[34] Weir fishermen capitalized on these commercial networks at the expense of upstream fishermen and the overall health of the Penobscot fishery. With the continued expansion of this commercial enterprise, in 1872 the U.S. Fish Commission mapped the considerable distribution of weirs and pound nets throughout the bay and lower river. From Winterport down the river to Islesboro, the map identifies approximately 160 weirs and pound nets scattered across the eastern and western shore of the lower river and upper bay.[35]

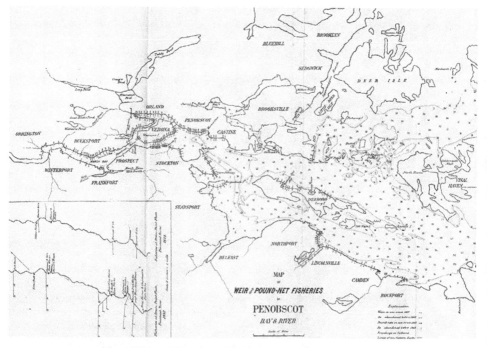

FIGURE 9. *Map of Weir and Pound-Net Fisheries of Penobscot Bay & River,* United States Commission of Fish and Fisheries, *Report of the Commissioner, 1872–1873,* Part II (Washington, Government Printing Office, 1874).

Between 1786 and 1814, Massachusetts attempted to rein in the commercial impacts with respect to the numerous sawmills and weirs that obstructed Maine rivers. Following precedent established by decades of inland fisheries regulation in New England, the legislature applied the same type of use limits to large nets, legal fishing days, and milldams.[36] With lawmakers situated in Boston, hundreds of miles south of these frontier waters, they delegated inspection and enforcement duties to committees formed by local townsfolk abutting the Penobscot River and waters "to the westward of." With the knowledge that "said fish do not usually go up some of the said rivers and streams and the branches thereof into the lakes and ponds to cast their spawn, so early as the first day of May annually," they trusted local forces to apply the appropriate regulations as migratory fish left the Gulf of Maine for freshwater systems.[37] Invested with this authority, these committees could authorize or discontinue the operation of milldams and weirs or require open fishways at their discretion. It is possible that fishermen and millers, displeased with the restrictions generated by a seemingly distant government, lobbied legislators for more regulatory discretion at the local level. Near the coast, local authorities understood that fishermen made their living from the water when soils failed to support a comfortable livelihood. Fishers and millers might expect local authorities to be far more sympathetic to the demands of commercial interests on the river, but the Massachusetts General Court continued to pass measures designed to moderate their commercial ambitions.

In 1810, lawmakers closed Maine's weir fisheries from July 1 to December 1 each year and required owners to install a four-foot opening to allow fish to pass on closed days during the fishing season.[38] State authorities had the difficult task of balancing the public good and the long-term health of inland fish stocks with the interests of commercial actors operating on tributaries and within the tidewater. Toward that end, by 1814, regulations compelled weir fishermen to seek a license from town selectmen to operate in the Penobscot. In addition, they had to post a bond of at least three hundred dollars that would forfeit to the town in the event of a violation. The law also prohibited salting fresh shad and alewives after June 10 and salmon after July 1.[39] This measure incentivized the consumption and sale of fresh fish in the local economy, but ice and other crude refrigeration methods would soon make this prohibition obsolete. Despite what appears to be a well-intentioned effort to respect both commercial interests within the tidewater and the rights of rural and Indigenous fishers further upriver, securing a license to operate a weir seemed to require little more than the capital to build and maintain the structure.

As an overwhelming commercial presence in the tidewater, weir fishermen refused to accept any responsibly for the sad state of the Penobscot River fishery. To obscure their own unsustainable operations, weir fishers from Frankfort, on the west bank of the river, blamed drift nets or set nets—a six-to seven-inch mesh net that extended up and down the entire water column—for the collapse of salmon and shad. They rejected wholesale the notion that their operations should be regulated to protect the fishery and hoped that "our equal right to fish with wears may not be taken from us to improve the rights of others."[40] Fishermen from Bucksport and Prospect denounced a prohibition on weirs after the Fourth of July as an affront to their revolutionary heritage, "that memorable day, on which our liberties begun, ends the rights of fishermen." They went even further, tapping into deep-seated colonial prejudices to advance their commercial agenda. They accused upriver critics of inciting "their red brethren" against them, entirely discounting the generational knowledge that informed the Penobscot Nation's warnings of environmental destruction on the river. Instead, weir fishers offered their own peculiar and ambiguous diagnosis for the loss of fish— what they termed "expert fishermen on the big waters" working in concert with "fish merchants"—but they concealed their own role in the commercial fishery.[41] With this belligerent posture, weir fishers failed to grasp, or willfully ignored, the ways their own operations infringed on the rights of upriver communities.

With a combination of environmental constraints and commercial opportunities, fishermen on the coast and within the estuary, on both the Penobscot and Kennebec, demonstrated an inclination to overexploit their fisheries. With soils ill-suited for agriculture, fishermen from Georgetown, near the mouth of the Kennebec River, looked to the "the river and cod fishery" as "the only resource whereby to gain a subsistence."[42] Some members of this coastal community sold barrels of alewives as bait for the lucrative offshore cod fleets. Like their counterparts on the Penobscot, they cautioned that upriver critics unfairly maligned their choice of fishing equipment. Instead of weirs, these fishermen asked fish wardens to inspect the large number of sawmills situated on streams and brooks near the tidewater.[43] Amid the dispute over where responsibility should lay for the destruction of migratory fish runs and the power of state government to intervene so as to reverse those trends, timber interests staked their claim to both rivers. In 1834, a timber firm installed the first dam on the Penobscot River. There was little, if any, recorded opposition to the dam. When corporate interests proposed to dam the Kennebec River the same year, arguments in favor of keeping the river in its natural state emerged from rival commercial sectors within Maine's coastal economy.

In 1834, Greenleaf White and a number of associates petitioned the Maine legislature to form a corporation to build a dam across the Kennebec River at Augusta. The Kennebec Mill Dam Association argued that this project would serve not only the interests of their corporation but also the population of the Kennebec River valley as a whole. They predicted that their dam would improve navigation for commercial vessels near Augusta and increase the value of the region's timberlands, with new mills "preparing it in the several modes required for the market." They articulated an ambitious vision for the future of the Kennebec River valley where "manufacturing establishments" harnessed both the water and the forest for industrial production.[44] As it turned out, the Edwards Dam would supply power to several paper mills in the Augusta area. Though the Kennebec Mill Dam Association worked hard to convince the general public that they stood to share the benefits of the initiative, many towns remained suspicious of this supposed corporate benevolence.

Property owners whose land abutted the Kennebec above and below the proposed dam immediately sensed the impending damage to their economic fortunes. Much like the Concord River farmers some thirty years earlier, they understood that their lands would be subjected to unpredictable flooding.[45] Waterville farmers spoke from experience and pointed to a particularly destructive flood in 1828 in which the river rose twenty inches within a few minutes due to an ice dam at Brown's Island.[46] They expected similar floods to accompany White's proposed dam. Farmers were convinced that their intimate experience with the environment would carry weight.

Sawmill owners objected to the Kennebec Dam on the grounds that it would impede the transportation of logs downriver. Loggers harvested timber deep in the interior of the Kennebec valley, tied the logs together in manageable units called "rafts," and floated them down to sawmills near the tidewater. Greenleaf White's proposal included the construction of a lock and canal system to enable commercial vessels to circumvent a series of natural obstructions near Augusta. This idea did not sit well with mill owners. Arthur Barry and other millers from Gardiner worried this project would hurt their bottom line since it would be impossible to fit rafts of "the customary dimensions" through the lock system. Increased transportation costs would affect the price of timber for both millers and shipbuilding firms that "rely upon this source for an important portion of the material necessary for carrying on their respective occupation."[47]

The coastal millers and fishers of Georgetown, who had previously accused each other of abusing the Kennebec fishery, found common ground in resisting the dam proposal at Augusta. Both of their economic fortunes were inextricably linked to the river. Residents of Georgetown wrote, "They and their fathers have from necessity been compelled to seek employment other than that of agriculture as their soil would not repay the expense of very general cultivation, they have been forced in a great measure to seek other means of subsistence."[48] Even though fish populations had already been seriously compromised, fishermen still presented a significant obstacle to the construction of a large dam across the Kennebec

River. Despite these declines, the dam proposal elicited the outrage of fishermen whose livelihoods depended on a reasonably healthy river fishery. Enough fish must have remained in the Kennebec to inspire hope that the fishery was not beyond salvation, otherwise stakeholders would have simply ceded the river to manufacturing interests without a fight.

The Edwards Dam threatened a multitude of commercial enterprises in northern New England, but as many complaints related, the overwhelming destruction wrought by obstructing the entire width of the Kennebec focused on fisheries. Fishing communities highlighted the broad economic implications associated with the dam proposal. Arthur Barry, the miller and fisherman from Gardiner, estimated the value of the Kennebec River fishery at "many thousand dollars . . . annually." The fisheries had for years "furnished employment to many industrious people" and constituted "one considerable item of the resources of the State."[49] In the words of Georgetown fishers, "The destruction of their fishing privileges would be infinitely more disastrous in its consequences—a large number of our citizens derive their only subsistence from the salmon, shad, alewife, and cod fishery." Without the fishery, the community may well be reduced to "positive beggary."[50] The feeling that a large dam at Augusta would prove disastrous for fish runs was no mere hypothesis. Petitioners from Woolwich, where Elias Robbins once fished Merrymeeting Bay, related their experience of the Androscoggin River in which dams "proved the destruction of a vast quantity of salmon which till then, had annually ascended that stream." They had no reason to doubt that the outcome would be any different on the Kennebec and argued that "should the Kennebec be obstructed by a dam, from shore to shore, we believe the various tribes of fish, visiting its waters from year to year, affording employment and sustenance to thousands of our fellow men, would be cut off at a stroke."[51] According to these petitioners, economic security for these riparian and coastal communities depended on free-flowing currents.

Alongside these economic arguments, petitioners emphasized the ecological systems that would be put in jeopardy with this industrial dam. Fishers and millers who had interacted with coastal and upstream ecosystems throughout their lives offered arguments predicated on local ecological understandings. Fishermen cured salmon and shad for urban markets, but alewives often supplied the coastal and offshore fisheries with a cheap source of bait. As a result, coastal fishers were attuned to the connection between river and marine ecology. They described how, in their experience, the cod fishery along the coast "depends entirely" on the seasonal migrations of alewives.[52] Codfish preyed on schools of alewives off the coast, which allowed residents an opportunity to fish for this demersal species without the significant expense and danger of operating offshore. Alewives occupied an important position within the marine food chain and served as a bridge between fresh and saltwater ecosystems. Fisheries biologists now recognize that both alewives and herring are instrumental in maintaining the Gulf of Maine's inshore cod stocks. When alewife runs collapsed across coastal Maine in the 1960s, cod populations moved offshore to chase other forage fish.[53] Modern scientific analysis confirms what these coastal townspeople had suspected for some time. Fishermen from Woolwich similarly acknowledged "that the success of our cod fisheries near the shore, depend[s] very much on the small fish that visit our rivers."[54] Having worked these waters for generations, fishermen understood the interconnected components of the coastal ecosystem and the broader ecological implications of main-stem industrial dams. In fact, the Edwards Dam would come to obstruct critical spawning habitat for alewives, and the effects would ripple throughout the watershed and into the inshore cod habitat.

Finally, petitioners pointed to the stark divide between small commercial producers and the capital investors who had initiated the dam proposal. They wrote, "After taking in to view the advantage of four men of capital, and rich corporations, to invest their funds . . . the great mass of our frontier population are of the lower class in point of wealth, and derive a valuable part of their support from their fisheries." This was not some local miller obstructing fish runs with a small milldam but, rather, wealthy individuals seeking to capitalize on the manufacturing boom and undermine the traditional economic landscape. They appealed to the support fishermen had historically enjoyed from state policymakers in protecting river fisheries, "the care also with which preceding legislatures have protected the fish—and in fact, by a grant to the petitioners, the <u>many</u> would be injured, whilst the <u>few</u> might be benefited."[55] Framed as an irreconcilable clash between public benefits and narrow corporate interests, state authorities now had to sort through this conflicting testimony and deliver a decision that would inevitably produce winners and losers. There is no doubt that this would be especially difficult in a climate of competing visions as to what exactly served the best interests of the public at large and not simply a small group of wealthy entrepreneurs.

The response of the Maine legislature to the objections of Kennebec River fishers, millers, and landowners reveals a shift in legislative and economic priorities during the era of industrial takeoff. Despite the tremendous resistance from small commercial producers across the Kennebec River valley, the legislature showed little interest in deviating from the industrial momentum that had already transformed southern New England and brought windfall profits to the investor class. On February 20, 1834, a special committee of the Senate and House of Representatives convened to review Greenleaf White's original petition to incorporate the Kennebec Mill Dam Company. They listed the major themes that defined community resistance: destruction of fisheries, the obstruction of log drives, potential damage to sawmills, and increased flooding near the dam. The committee dismissed these objections as "not sufficiently valid," with no justification as to how they arrived at that conclusion. The committee ultimately invited White and his associated to introduce a bill that granted incorporation and permission to construct a dam in Augusta.[56] When they finished construction in 1837, contemporaries estimated that Augusta was destined to become one of the largest manufacturing cities in the entire nation.[57]

In the end, damage to commercial fishers, sawmill operators, and shipbuilders failed to convince state authorities to hold off the installations of large dams in Maine. The allure of transforming Augusta into a profitable manufacturing city on par with Lowell or Lawrence was simply too good to dismiss. This may have also been a case of too little too late, as fishermen had failed to negotiate and compromise on regulations designed to check overfishing in the previous decade. Nevertheless, even a coalition of farmers, fishers, and millers could only forestall the inevitable for so long. While fishers, farmers, and millers were able to preempt the arrival of industrial dams on the Merrimack in 1801 and 1812, they, too, would succumb to the momentum of textile manufacturing and the promise of national economic development.

Closing the Merrimack

Francis Cabot Lowell and the Boston Associates formed the Boston Manufacturing Company in 1813. They quickly set up shop on the Charles River in Waltham, Massachusetts, and revolutionized textile manufacturing with their new technically sophisticated, integrated production system. Lowell died in 1817, but the associates pushed forward with plans to expand their operations to the nearby Merrimack River. Incorporated in 1822 as the Merrimack Manufacturing Company, "for the purpose of manufacturing and printing cotton goods at Chelmsford in the county Middlesex," Kirk Boott and Nathan Appleton transformed this agricultural backwater into the manufacturing city of Lowell.[58] In 1826, the associates acquired the Proprietors of Locks and Canals on the Merrimack River and all the accompanying land and water rights, including the Pawtucket Canal, thus launching a new era of industrial manufacturing in North America. According to Theodore Steinberg, "The marketing of waterpower at Lowell and the other Merrimack Valley towns embodied a relationship with nature that defined water principally in terms of its value for exchange and production."[59] With water now characterized as a commodity, the Boston Associates relied on scientific expertise to obtain all the power the Merrimack River could offer. New Englanders anticipated another conflict over the right to retain access to the Merrimack River fishery, with the waters below Pawtucket Falls offering one of the most fruitful sites for textile production in New England.

By the 1820s the Merrimack River fishery was suffering from a combination of industrialization and overfishing. Textile mills at Lowell began turning out finished cloth in 1823, utilizing a strategically placed dam amid the existing canal to generate power for the numerous spinning looms.[60] The writing was on the wall. No longer would the General Court protect the rights of farmer-fishermen at the expense of industrial manufacturing, and yet industrial manufacturing was not entirely responsible for the wholesale abandonment of fisheries conservation on the Merrimack River. In the midst of this industrial transformation, the General Court accepted petitions from fishermen and millers who wished to operate free from regulation. In 1834, a local mill operator named John Tenney petitioned the government to repeal existing regulations that allowed for the passage of fish around mill sites. He clearly spelled out the useless nature of fisheries regulations, as the "quantity of alewives taken in this brook are of small and inconsiderable value." Further, "the time spent looking after fish is of much more value than the fish themselves, and could be more usefully employed if there were no fish in said brook to draw attention toward them."[61] With fish populations now a far cry from the abundance detailed in George Brown Goode's fisheries report, farmer-fishermen were forced to come to terms with a harsh reality. Because there was no longer a healthy fishery to protect, the General Court granted this petition and allowed the mill to operate free of the restrictions put in place by previous regulations, a sign of things to come in regard to the Merrimack River fishery.

In 1846, fishermen from the towns of Bradford and West Newbury—where the battle to save the Merrimack fishery had commenced—petitioned the legislature to repeal all fishing regulations on the Merrimack River, declaring that the existing laws "regulating the Fishery in Merrimack River are to a great extent a dead letter."[62] With the migratory fish limited to well below the great dams constructed at Lowell and Lawrence, fishermen wished to ply their trade unencumbered by laws written and enforced before the erection of these impassable structures. The General Court granted this request, which represented not only the end of migratory fish in the Merrimack River but also the end of the Merrimack valley farmer-fishermen. They had vigorously defended their traditional right to fish the waters of the Merrimack River, a rich ecosystem, since "time immemorial," but there was nothing left to defend—the fish had disappeared. River fisheries remained productive further north in the state of Maine for some time, but on most main-stem channels, fisheries preservation no longer represented an impediment to industrialization.

In the Northeast, the transformation of freshwater ecosystems began with canals and concluded with large industrial dams. Rural coalitions on the Merrimack River in the early nineteenth century could only forestall an industrial takeover for so long, and stakeholders further north in Maine failed to unite behind a cohesive fisheries conservation program. Extensive divisions in the Penobscot and Kennebec fishing communities made the already difficult task of responding effectively to industrial interests nearly impossible. In the absence of productive soils, they relied heavily on fish resources to cobble together a living and therefore resisted most forms of regulation. Maine stands as an example in which fragmented fishing interests and overreliance on fish resources undermined a unified resistance to an advancing industrial agenda. Maine river fishers consistently missed opportunities to unite and advocate for regulations that would benefit all and instead demonstrated far more inclination toward the commercialization of natural resources.

By 1834, much of the Northeast had been transformed by the new manufacturing economy, and capitalists sought to replicate that success wherever rivers offered sufficient power. But because farmer-fishermen had fought to preserve this resource for over a century before industry advanced its own claim to inland waters, fishermen and state authorities began to explore alternative strategies such as fish ladders, open sluice ways, or other forms of remediation to partially restore degraded fisheries.

The search for the appropriate balance of regulation was ultimately cut short by the appearance of industrial dams, and yet it remains important to note that some communities, at least up to this point, felt they possessed the knowledge and ability, with the aid of sympathetic legislators, to manage river fisheries for sustainable returns. Long before industrial manufacturers staked their claim to the Merrimack River, rural communities argued for sustainable fishing, campaigned against those who sought to monopolize the catch, and laid the groundwork for renewed interest in the economic and ecological benefits that accompany the revival of native anadromous species. With the onset of industrialization, this feeling dissipated, leaving only a few individuals such as Thoreau to mourn their disappearance and hope that "after a few thousand of years, if the fishes will be patient, and pass their summers elsewhere, meanwhile, nature will have leveled the Billerica dam, and the Lowell factories."[63]

As it happens, Thoreau would not have to wait quite that long. A remarkable turnaround in the last several decades has shifted this narrative once again in favor of the fish. Coalitions of local stakeholders, Indigenous communities, environmental organizations, and state and federal agencies have partnered to remove dams and restore native aquatic species. In scientific circles, and increasingly accepted by the general public, healthy freshwater ecosystems are linked to biodiversity and the return of native species. As such, anadromous fish are beginning to reclaim their position as symbols of clean and healthy rivers.

Never missing a chance to highlight American failings and extol good British governance, the English gentry and sport fishermen of the mid-nineteenth century concluded that "salmon flourish under the cold shade of aristocracy; democracy, as in Canada, or the United States, is death to him."[64] Although democracy once played a part in the decline of river ecology, over the past several decades networks of conservation groups, resource managers, Indigenous people, and local stakeholders are pushing for new environmental policies. Aristocratic finger wagging aside, it does appear that democratic, grassroots energy can equally restore degraded environments. More contemporary examples of river restoration bear out such a seemingly optimistic vision for the future of New England's rivers.

CHAPTER SIX

RIVERS RESTORED

Almost immediately after the new industrial economy had extended its reach to nearly every major waterway in New England, state programs aimed at saving the fast-disappearing native sea-run fish emerged to counterbalance widespread enthusiasm for dams, technological innovation, and the promise of shared prosperity. Corporate interests had applied scientific and technical expertise to manipulate and harness flowing water to power industrial manufacturing, but newly formed state fish commissions hoped that science and technology might also mitigate the worst effects for New England's celebrated seasonal fish runs. The work commenced in Vermont when towns along the Connecticut River petitioned the state for some coordinated plan to address the destruction of the salmon and shad fishery. In 1857, the governor appointed George Perkins Marsh as Vermont's first fish commissioner and asked that he develop a set of proposals that might restore native species to their historic habitat.[1] Marsh was uniquely qualified for this difficult assignment. Though he enjoyed a distinguished career as a lawyer, scholar, member of the House of Representatives, and foreign minister to Turkey and the Kingdom of Italy, as a public intellectual Marsh challenged popular assumptions of humanity's supposed benign relationship with the natural world. In his 1864 environmental classic, *Man and Nature*, Marsh observed that "the ravages committed by man subvert the relations and destroy the balance which nature had established between her organized and her inorganic creation."[2] Attuned to the ecological connections that sustained both natural systems and the communities that depended on them, Marsh articulated a conservation vision to counteract reckless, and potentially irreversible, environmental damage. Nearly a decade before the publication of *Man and Nature*, in his "Report on the Artificial Propagation of Fish," Marsh applied this revolutionary environmental perspective to New England's freshwater ecosystems.

During Marsh's lifetime, Vermont's rivers, streams, and lakes had experienced steady declines in fish populations, with earlier efforts to reverse this unwelcome trend and restock fish to significant numbers failing to yield meaningful results. He wrote, "It is well known that in the earlier periods of the history of Vermont, the abundance of fish in the running waters, and more especially in the ponds and lakes of our interior and our borders . . . Lake Champlain and the Connecticut, as well as those of their larger tributaries whose course was not obstructed by cascades, abounded in salmon."[3] To recover some portion of that historic abundance, Marsh called for interstate compacts to mitigate overfishing in the tidewater, curb industrial pollution, and pass fish around large dam obstructions, but he also applied a more holistic reading of the fisheries problem that, according to biographer David Lowenthal, "showed him alert to the interplay of plant and animal habitats, and alarmed by the human impairment of their intricate linkages."[4] For Marsh, forests, watersheds, and soils represented interdependent parts of a greater whole. Changes in the land would inevitably spill over into the water. Most concerning was the rapid pace of deforestation that Marsh correctly observed accelerated soil erosion, increased flooding, filled streams with sediment that destroyed spawning habitat, increased water temperature, and disrupted the aquatic food chain. With this unique ecological sensibility, Marsh added deforestation to the growing list of obstacles to the restoration of migratory fish but also broadened the conservation mission to include the state's forests and soils.[5]

Since deforestation, industrialization, and commercial overexploitation had so compromised the ecological integrity of the region's freshwater ecosystems, Marsh and his counterparts embraced fish culture, the artificial breeding and stocking of native species, to recover populations of sea-run fish like salmon and shad. On interstate waters like the Connecticut River, this would require the cooperation of legislators, bureaucrats, scientists, and private citizens from Connecticut, Massachusetts, New Hampshire, and Vermont, but artificial propagation also demanded that fishermen come to terms with the fact that, according to Marsh, "we cannot destroy our dams, or provide artificial water-ways for the migration of fish, which shall fully supply the place of the natural channels; we cannot wholly prevent the discharge of deleterious substances from our industrial establishments into our running waters." Given the constraints imposed by the industrial takeover of the region's large main-stem rivers, Marsh believed it was also unlikely that "any mere protective legislation, however faithfully obeyed, would restore the ancient abundance of our public fisheries."[6] Yet faith that artificial propagation might recapture a "public fishery" would prove equally misguided. By the 1870s, state fish commissions aggressively pursued restocking programs, oversaw the installation of fishways around impassable dams, and drafted laws to restrict fishing; nevertheless, migratory fish still found it difficult to navigate the aquatic labyrinth of concrete obstructions, chemical pollutants, and rogue fishermen. All the public support, significant financial outlays, and technical expertise could not replicate clean, oxygen-rich habitats and free-flowing corridors for upstream migration. What had been lost could not be so easily recovered. The river commons, a "public fishery" in Marsh's terms, collapsed under the weight of commercial and industrial capitalism. Still, powerful memories of seasonal abundance brought rural communities, politicians, and bureaucrats together to begin the process of reintroducing native species to their historic habitat.

Mid-nineteenth-century efforts to restore New England's depleted rivers were pushed forward by two hundred years of consensus in the countryside testifying to the cultural and economic significance of seasonal fishing. Though Marsh's work would help popularize what environmental circles today refer to as "ecological restoration," the roots of inland fish conservation emerged from rural communities that fought to defend these resources from commercial abuse and industrial takeover. In 1856, many farmer-fishermen remembered well the productive seasonal fish runs that filled their tables and contributed to a vibrant rural economy. In 1865, residents of Chelmsford, Massachusetts, who had witnessed firsthand the ecological repercussions of large dams, industrial pollution, and overfishing, estimated that the Merrimack River had once supplied a third of all animal food consumed in the town. As such, petitioners from Massachusetts and New Hampshire joined Vermont fishermen in calling for new measures to facilitate the revival of anadromous species.[7] State fish commissioners were simply responding to the will of the people in putting forward a comprehensive conservation agenda aimed at the revival of this important resource. With industrial dams powering the manufacturing revolution, fish commissioners sought to identify common ground where fisheries and industry might coexist. If fisheries had not held some measure of significance within the historical memory of these rural communities, state authorities would not have expended any effort contemplating this middle path.

Restoration Gains Traction

Throughout the 1860s, New England led the way in establishing the first permanent fish commissions in the United States. These commissions responded to a profound sense of discontent among rural farmer-fishermen who were keenly aware of what had been lost. In 1866, fish commissioners for the Commonwealth of Massachusetts, Theodore Lyman and Alfred A. Reed, released their *Report to the Senate concerning the Obstructions to the Passage of Fish in the Connecticut and Merrimack Rivers*, which laid out moderate proposals revolving around the installation of fishways at major dams that they hoped would, in some measure, restore migratory fish populations to sustainable levels. Born into an elite Boston family with significant investments in manufacturing on the Merrimack and Connecticut Rivers, Lyman seemed an odd choice for this position; however, with a background in science and public service—and a passion for sport fishing—he believed in the power of technical expertise to solve the problem of the commonwealth's depleted rivers.[8]

Lyman and Reed approached their work with a clear-eyed sense of the challenges that lay ahead. Their report described the ways in which industry had refashioned nature to meet the energy needs of the new industrial economy, "such is the history of the Merrimack, and its great source Lake Winnipiseogee [sic], whose waters, once free, now possess only a conditional right of way and work their painful passage to the ocean by turning a thousand wheels."[9] The river had been subdued—made tame and predictable by storage dams near the headwaters that delivered an even supply of water during times of drought or flooding. Further south, additional dams and canals fed water to a series of turbines, gears, belts, and wheels that supplied the mechanical energy necessary to power spinning looms. Lyman and Reed understood that migratory fish would require assistance in navigating these new bionic rivers and proposed a $22,000 investment in three large fishways to allow fish to surmount large dams on the Connecticut and Merrimack Rivers.[10] By the end of the year, the Massachusetts Fish Commission released plans for new, and hopefully more effective, fishways at Lowell and Lawrence.[11]

Fish commissioners also made an effort to protect fish stocks in southern New England, where rapacious fishermen targeted migratory species below impassable dams. Beginning in the 1866, southern New England states instituted laws to protect salmon in an effort to provide safe passage to upland spawning grounds.[12] Next, Massachusetts and Connecticut both passed laws that required passageways around large main-stem dams that obstructed the lower reaches of the Connecticut and Merrimack Rivers. Early in this experiment with fish ladders, commissioners reported that fishways on the Merrimack River functioned reasonably well.[13] The last piece of the puzzle, fish commissioners in New Hampshire and Vermont stocked the headwaters of both the Merrimack and the Connecticut Rivers with salmon fry from the Miramichi River in New Brunswick, Canada, and, by 1871, the Penobscot River in Maine. In fact, Maine fish commissioner Charles Atkins pioneered a system for breeding salmon that has continued for 125 years.[14] The Craig Brook National Fish Hatchery in Orland, Maine, gained such a positive reputation that fish hatcheries across the United States acquired salmon eggs from Craig Brook.

For midcentury fish commissions the path forward was clear. Successful fisheries restoration hinged on a multifaceted program devoted to strong regulations, technical modifications to impassable dams, and artificial propagation. A spirit of optimism pervaded these conservation initiatives. State fish commissions operated with strong faith in the power of scientific and technical expertise to overcome the challenges of resource depletion, the same ideology that would later define Progressive Era conservation. In hindsight, perhaps this optimism was misplaced.

As fish populations began to rebound, nearsighted fishermen continued to harvest anadromous species despite a tenuous recovery. State laws that protected the slowly recovering fish runs proved difficult to enforce. Commercial shad fishing continued nearly uninterrupted below the Essex Dam on the Merrimack. In 1841, fishermen took approximately 365,000 shad in the lower portion of the river. Over the next thirty years, the figure plummeted to 11,255.[15] Growing dissatisfaction among commercial fishers led Connecticut to repeal regulations protecting salmon in the 1870s. During the 1890s, New Hampshire commissioners loosened their oversight of fishways and illegal fishing.[16] Following Connecticut's lead, in 1895 Massachusetts abandoned all fishing restrictions within the tidewater of the Merrimack River.[17] Despite early success in passing fish around dams and the large-scale restocking programs that released as many as one million shad eggs into the Merrimack in 1867, the initial spirit of interstate cooperation that colored midcentury fish conservation could not be sustained.[18]

FIGURE 10. *View of the Falls at Lawrence,* 1893. Essex Dam in Lawrence, Massachusetts. Courtesy of the Lawrence Public Library.

Part of the problem rested with inconsistent compliance from dam
owners. State authorities failed to compel factory owners to maintain the fishways that state authorities engineered for their dams.
Discouraged with rising costs, slow results, and increasing levels of
industrial pollution, New Hampshire abandoned all fishway maintenance in 1890. When a flood damaged the fish ladder at the Essex
Dam in 1896, the replacement proved ineffectual, and manufacturers allowed it to fall into disrepair.[19] Though mill owners at Lowell and Lawrence on the Merrimack largely complied with direc-
tives to maintain passageways, manufacturing interests were not of
one mind on this issue. The Holyoke Waterpower Company on the

Connecticut River refused outright to construct a fish passage and took the fight all the way to the Supreme Court. In 1872, the court ruled in favor of Theodore Lyman and the Massachusetts Commission on Inland Fisheries, but this confrontation did not bode well for the future of industrial cooperation with state fisheries conservation.[20]

Inland fish conservation proceeded in fits and starts and never secured the type of long-term political commitment to interstate programs that had clearly proven to bear fruit. From 1877 to 1898, fisheries scientists estimated more than twenty thousand salmon had passed through the Essex Dam fishway in Lawrence.[21] These numbers would have likely been even greater if state authorities continued to restrict fishing below the dam and within the tidewater of the Merrimack. The blueprint for success was clear. As Lyman and Reed affirmed in their report to the Massachusetts legislature, "The best contrived fish-passages, and the most extensive breeding of the young fry, would avail little without proper laws rigidly to regulate the time and manner of fishing." This three-pronged approach set an important precedent for those who remained committed to the cause of inland fisheries conservation. States continued to follow these conservation strategies, with varying degrees of commitment, into the early twentieth century, but the goal of self-sustaining, seasonal fish runs remained elusive.[22]

Since state fish commissions failed to deliver on the promise of restoration, the fisheries conservation agenda shifted upland to remote lakes, ponds, and streams where the conditions could be managed with far more precision. Within these comparatively pristine waters, relatively insulated from industrial or commercial interference, fisheries managers catered to a new class of fishermen: genteel sportsmen that visited northern New England to escape the trappings of urban life and connect with the masculine traits popularly attributed to outdoor recreation. Starting in the late nineteenth century, conservation rights began to privilege wealthy sporting interests—anglers and trophy hunters—and an emerging tourist industry that promised much-needed economic development for depressed rural communities. According to Richard Judd, from 1865 to 1900, fish and game clubs and resort interests "linked arms with state fish and game commissioners and helped turn the conservation movement toward urban formulations of common-resource use."[23] This new conservation constituency composed of outdoor recreation and tourist industry interests often clashed with locals that held on to rural traditions of economic independence and livelihoods organized around semisubsistence production.[24] Gone were the days in which rural farmer-fishermen alone held common rights to local waters for household use and community trade. While Progressive Era sporting interests focused their attention on protecting clean backcountry trout waters, social reformers and game managers partnered to address growing concerns that urban and industrial pollution on large watersheds threatened both aquatic life and the public health.

By 1870, the owners of textile, paper, leather, and metallurgy companies set out to redefine rivers primarily as industrial drains—siphoning away the odious by-products of the new modern economy. Dissolved oxygen levels, necessary for sustaining all aquatic life, fell precipitously as industry dumped high concentrations of nitrates and phosphates into the region's inland waters. Reformers pushed for state intervention, and in 1869 Massachusetts paved the way by establishing the region's first board of health. Over the next two decades, the rest of New England followed Massachusetts's lead in confronting the harmful effects of water and air pollution. State agencies attempted to mitigate the worst effects of freshwater pollution but consistently encountered a fundamental tension between public health, communal rights, and corporate interests.[25] For much of the twentieth century, large watersheds were so polluted that fisheries restoration, let alone conservation, was a futile exercise. It was not until 1960s that river restoration would once again capture the public imagination thanks to federal policies that put rivers back on the path to sustaining a broad array of aquatic life.

Resilience and Recovery

Historian Samuel P. Hays argues that Progressive Era conservation consciousness shifted after World War II and led to the birth of the environmental movement. Postwar economic growth supported the emergence of a strong middle class that prioritized clean and healthy living spaces as a foundation for new standards of living. This new movement focused on issues of public health and access to wilderness in an increasingly polluted and urban society.[26] The movement to restore healthy urban environments began with the low-hanging fruit of polluted rivers. Rivers that traveled through industrial centers in Pittsburgh, Cleveland, Chicago, New York, and Boston demanded the public's attention. Cleveland's Cuyahoga River famously caught fire in 1969 as result of decades of chemical pollution. This event helped crystallize the environmental movement and center river restoration at the core of the public consciousness. Environmental organizations worked to redefine rivers as a public trust, a resource for all citizens. The work of environmental activists paid off considerably with one of the era's landmark legislative achievements, the Clean Water Act. Passed in 1972, the law ensured that rivers no longer exclusively catered to an industrial and chemical constituency.

The Clean Water Act brought about dramatic improvements in water quality. One example in Maine showed that with the necessary political will and public support, dramatic reversals in ecosystem health were within reach. Throughout the 1960s, the Kennebec River was so polluted that fumes from the river peeled paint from nearby homes. From 1972 to 1990, the Clean Water Act provided funding for sewage treatment infrastructure and imposed federal guidelines for pollution control that yielded a 95 percent improvement in water quality.[27] No one thought the restoration of migratory fish species a remote possibility before this radical transformation. As the river slowly came back to life, environmental advocates focused their attention on an element of the ecosystem that had long been forgotten.

Some scientists argue that North America is still home to the greatest source of temperate, freshwater biodiversity on the planet. Despite this highly prized designation, the continent also hosts some of the most degraded and threatened freshwater habitats in the world. Within the last thirty years, the number of endangered freshwater species has increased dramatically.[28] Diadromous species have been hit particularly hard, with numbers throughout the North Atlantic at historic lows.[29] In Maine, home to some of the last viable diadromous fish runs, both the Atlantic salmon and shortnose sturgeon are currently listed under the Endangered Species Act. American shad are now found in only five rivers throughout the state.[30] Because the State of Maine historically hosted some of the healthiest stocks of diadromous species, the transformation from abundance to scarcity has been especially stark. The root cause is apparent even to the casual observer. Loss of habitat, overfishing, and pollution combine to bring about depletion and, in some unfortunate cases, extinction. However, the casual observer may not understand that these challenges have deep roots in New England's environmental history.

The march of industrial and urban development made an unmistakable imprint on the landscape. In 2009, estimates pointed to some eighty thousand dams larger than six feet tall scattered throughout the United States. The total is actually closer to two million. Nearly every large river today hosts at least one significant dam.[31] Dams no longer power American manufacturing directly, but many still serve a purpose. They create reservoirs for drinking water, irrigation, recreation, and flood management, but energy production is still the most common rationale for relicensing large dams. Hydropower dams generate 7 percent of the nation's energy demands, not an insignificant figure, and represents one of the largest sources of renewable energy in the United States.[32] While some environmental circles welcome the growth of hydropower as an alternative to fossil fuels, large dams extract significant environmental costs. They fragment otherwise connected ecosystems, block the flow of nutrients and sediment, drown floodplains beneath massive reservoirs, and, of course, destroy fish and wildlife habitat.[33] More recently, climate change has created an entirely new set of challenges for freshwater ecosystems. Warming temperatures have accelerated the timing of spawning runs, with migratory fish appearing in freshwater ecosystems earlier than in the past. The ecological effects of these phenomena have yet to be determined, but most scientists agree that any disruption in the delicate ecological balance that supports diadromous species across two complex, yet interconnected, ecosystems—marine and freshwater—is unlikely to end with positive results.[34]

Despite this bleak picture of environment degradation across two hundred years of urban and industrial expansion, there is cause for hope. Though migratory fish populations have been reduced to historic lows, extinctions are rare. Instead, biologists have used the term "ghost species" to describe the precarious state of North Atlantic diadromous species. John Waldman, a prominent conservation biologist and an important voice in this debate, took this metaphor one step further in laying out the difficult work that lies ahead: "The task today is to exorcise these ghosts, not through the supernatural but by filling the empty spaces in nature they represent through the hard work of applied restoration via all possible avenues."[35] The task of replacing these empty spaces with native diadromous species has represented the guiding light of river restoration. Contemporary restoration efforts proceed from an understanding that biodiversity represents a defining feature of healthy and stable ecosystems.[36] Since large dams have discouraged the return of native species and have thus disrupted systems of interaction between species, many have come to question the wisdom of maintaining them. But it bears repeating that suspicion of dams as a disrupting force for ecological stability is not a new phenomenon. Up until the mid-nineteenth century, large coalitions of rural New Englanders opposed dam construction and the resulting degradation of migratory fish habitat. It is not particularly difficult to see the ideological connections that link preindustrial coalitions to the work of present-day conservationists. At the foundation of these traditions is a shared commitment across generations to the maintenance of healthy ecosystems that support river fisheries.

Beginning in 1999, the removal of the Edwards Dam on the Kennebec River rejuvenated the debate surrounding the ecological impact of dams and the historical significance of the river's fisheries. Just as proposals for the construction of the Edwards Dam in the nineteenth century provoked tremendous controversy among rival commercial stakeholders, calls for its removal set up a familiar confrontation between corporate actors and grassroots activists. When the dam's license came up for renewal in 1993, a coalition of local and national environmental groups, state authorities, and federal agencies stood in opposition. Improvements to the river's health made possible through Clean Water Act investments and regulations sparked hope that the river could once again host a broad range of aquatic life. Local outdoor recreation enthusiasts and environmentalists alike were emboldened as biologists introduced a plan to restore native species of fish in 1985.

When federal energy regulators took up the issue, the Edwards Manufacturing Company most likely assumed that economic considerations would override the concerns of environmental organizations, as they had for over a century. However, in a decision that came as surprise to both sides, the Federal Energy Regulatory Commission (FERC) refused to issue a new license and called for the removal of the dam. Regulators concluded that the environmental costs outweighed the value of relicensing the Edwards Dam. The dam itself generated only a modest three-and-a-half megawatts of electricity, and new protocols required that the FERC balance energy production with environmental impacts in the decision-making process.[37] The removal of the Edwards Dam made history as the first time that regulators approved the destruction of a dam against the wishes of the owners and set an important precedent by which concerns for the health of the river's fisheries once again held weight with policymakers.

On July 1, 1999, the dam came down, and seventeen miles of the Kennebec opened to migratory fish for the first time in over a century. In 2008, state officials demolished the Fort Halifax Dam in Winslow, Maine, which opened additional spawning habitat on the

Sebasticook River, a tributary of the Kennebec. As a result, water quality between Waterville and Augusta improved dramatically. Consequently, the Kennebec watershed now hosts the largest run of alewives and other river herring on the entire Eastern Seaboard. In 2014, the Maine Department of Marine Resource recorded more than two million alewives at Benton Falls on the Sebasticook River. Shad have also been migrating in greater number upriver from Augusta, where the Edwards Dam once blocked their path. Given this remarkable turnaround, local newspapers characterized the river restoration project as a "wildlife success story" that was "breathing new life into the river."[38] Encouraged by this achievement, the same year the Edwards Dam came down, a campaign further north to rid the Penobscot River of several dams reached a critical juncture.

A New Mainstream

The pace of dam removal has stepped up dramatically since the turn of the twenty-first century. In 2015, *Science* magazine celebrated the removal of over one thousand dams since the 1970s. More than half of the removals have occurred within the last ten years.[39] The progress of dam removal was also featured in the pages of *National Geographic*. In an article commemorating the world's largest dam removal—the 210-foot-high Glines Canyon Dam on the Elwha River in northwestern Washington—Bob Irvin, president of the conservation organization American Rivers, describes the shifts in environmental attitudes brought about by these restoration success stories. He notes that thirty years ago the removal of the Glines Canyon and Elwha Dams would have seemed "a crazy, wild-eyed idea." But with more than one hundred dams torn down between 2012 and 2013, dam removal had become "an accepted way to restore a river. It's become a mainstream idea."[40] The Lower Elwha Klallam Tribe and environmental activists had long sought the removal of these two aging concrete behemoths, one of which, the Glines Canyon Dam, was actually located inside the boundary of Olympic National Park. In 1992,

Congress appropriated funds for the purchase of the Glines Canyon and Elwha Dams, and activists and Indigenous peoples quickly began building a local consensus for their removal.[41] Thanks to their work, the river's journey back to its natural state commenced in 2011. Now, the Elwha flows free, and salmon are beginning to reclaim to their historic habitat. Only time will tell as to what numbers of chinook, coho, chum, and pink salmon will rediscover these waters.

In some cases, these concrete barriers have stood in place for centuries and seem to many a permanent fixture of the landscape, but this has not prevented them from coming under intense scrutiny. Though many cannot imagine nearby rivers without the familiar sight of an artificial obstruction, a growing contingent of conservation-minded citizens have undermined the historical position of dams as immutable features of the land and water. Two monumental projects on opposite ends of the nation are largely responsible for this revolutionary perspective and have brought dam removal to the forefront of mainstream environmental consciousness. The revitalization of the Penobscot and the Elwha Rivers has not only proven the concept of river restoration through dam removal but also demonstrated that rivers possess a remarkable capacity to rebound from centuries of disconnection.

No project has helped bring dam removal into the mainstream for New England and the Northeast more than the recent transformation of the Penobscot River. Despite widespread support and enthusiasm for the removal of old dams, success was never guaranteed. Karen Limburg and John Waldman note that when "demand dwindles and constituencies are lost, it becomes increasingly difficult to motivate and secure funding for adequate management and restoration efforts."[42] As environments deteriorate over long periods of time, collective memory of ecosystems rich in biodiversity fades and broad interest in restoration tends to evaporate. Thanks to an unprecedented collaboration, the Penobscot Indian Nation and several nonprofit organizations—including American Rivers, the Atlantic Salmon Federation, Maine Audubon, the Natural Resources Council of Maine, the Nature Conservancy, and Trout

Unlimited—worked closely with federal and state agencies to "rebalance fisheries restoration with hydropower in the largest watershed within Maine," in the process negotiating the most ambitious dam removal project to date.[43] The national coordinator for the U.S. Fish and Wildlife Service's fish passage program, Susan Wells, described the scope of the project for a local newspaper: "When it is all said and done, it will be the largest river restoration project in the country."[44]

Opportunity struck in 1999 when the PPL Corporation of Pennsylvania purchased a series of dams on the lower Penobscot River. Prior to the transfer, the Penobscot Nation worked with local and national environmental groups to contest the renewal of federal licenses that would keep these dams in operation for decades to come.[45] According to Laura Rose Day, the executive director of the nonprofit spearheading the resistance, the Penobscot River Trust, "The backdrop for the negotiations was total rancor on the river, over every individual dam relicensing."[46] To avoid years of legal challenges, the restoration coalition and PPL agreed to pursue a comprehensive agreement for the future of the Penobscot River. Signed on June 25, 2004, the Lower Penobscot River Settlement Accord outlined a path toward a connected ecosystem, opening the river to Penobscot Bay and the Gulf of Maine. The Penobscot Trust raised $25,000,000 to purchase three dams—the Great Works, Veazie, and Howland Dams—on the lower Penobscot watershed. Both parties agreed to remove both the Great Works and the Veazie Dams and construct a bypass around the Howland Dam further upriver. In addition, the company agreed to construct and maintain a state-of-the-art fish lift at the sole remaining hydropower dam on the lower Penobscot River. In return for these concessions, the trust agreed not to contest hydropower licenses for six dams located on tributaries of the Penobscot. The company then upgraded those sites to offset the loss of hydropower previously generated by the three decommissioned dams on the main-stem river.[47]

At the ceremony marking the breaching of the Great Works Dam in 2012, Secretary of the Interior Ken Salazar remarked that the Penobscot Restoration Project represented "an important milestone for river conservation in America."[48] The restoration coalition celebrated the restoration efforts as "a model for cooperative conservation." Members did not overstate the gravity of what they managed to accomplish. The following year, water spilled over the concrete rubble of the collapsed Veazie Dam for the first time in more than one hundred years. Penobscot tribal chief Kirk Francis commemorated the historic occasion: "To the Penobscot, this river is our very soul. It's a place where we truly hold hands with our history and our ancestors and with each step of this project, we feel that much closer to it and realizing our people's dream of this sacred river in its natural state providing for and nurturing our people in many ways as it has always done."[49] After centuries of disconnection from river traditions that predated colonization, Francis communicated the hope that some measure of the tribe's traditional relationship with the river might be restored alongside the native species of migratory fish. Frank Speck, an anthropologist who studied Penobscot culture in the early twentieth century, wrote that salmon of "unheard-of quantities were taken here by the tribe each year until the dam was built. In those days they feasted on the fresh fish and smoked a large amount of it for winter upon pole racks over a fire."[50] As one of the primary stakeholders, the Penobscot Nation celebrated the restoration project as the culmination of a generations-long struggle to reconnect to their ancestral fishing traditions and with their riparian homeland. Should native fish populations rebound, the hope is that the Penobscot will once again resume subsistence fishing rights on the river that bears their name. With the anticipated return of Atlantic salmon to one of the most historically productive habitats in the Northeast, only time will tell whether the Penobscot might once again host significant populations of this celebrated species.

After the demolition of the Veazie Dam in 2013, the installation of the fish lift at the Milford Dam in 2014, and finally the dug-out bypass around the Howland Dam in 2015, over one thousand miles of habitat opened to eleven species of native diadromous fish. The lower river species such as Atlantic and shortnose sturgeon now had access to 100 percent of their historic habitat, while Atlantic salmon and other upper-river species benefited from those additional thousand miles of unobstructed river. In addition, the project allowed for new recreational opportunities, including paddling, fishing, and wildlife observation, accomplished without sacrificing a single watt of hydropower generation.[51] As fossil fuels continue to fall out of favor owing to climate change, there is an understandable reticence to abandon wholesale one of the nation's largest source of renewable energy. The Penobscot River Restoration Project has shown that both migratory fish and energy production can coexist.

Shortly after the restoration program concluded in 2015, the river experienced a dramatic turnaround and a resurgence of native fish populations. After completion of the Milford fishway in 2010, fisheries biologists began stocking alewives in nearby Blackman Stream and Chemo Pond. In 2015, more than half a million alewives passed the Milford fish lift on their way upriver to spawning grounds at Chemo Pond, tripling the previous year's count. The same year, more than eighteen hundred shad passed upriver, a dramatic improvement from the paltry annual figure of twenty fish in the previous thirty years. Once a ghost species, this historically significant fish has achieved an astounding recovery. Fisheries biologists have fielded inquiries about the best way to catch shad, which attests to how the fish have returned in the public consciousness after generations of scarcity. With the river flowing free once more, new sections of whitewater rapids brought paddlers throughout the region to the lower Penobscot. In 2015, the Penobscot River Whitewater National Regatta, held by the American Canoe Association, hosted more than one hundred avid paddlers eager to experience these recently unleashed currents, which was exactly what proponents of river conservation had in mind. The *Portland Press Herald* marked the two-year anniversary of the dam removals with a vision for the future of New England's rivers: "The goal is clear: Restore the health of river ecosystems so that fish populations can flourish again, and provide economic benefits that flow to local communities from healthy fisheries, paddling and other forms of recreation." [52] Between sport fishing and paddling, outdoor recreation has brought new economic opportunities for the Kennebec and Penobscot valleys. Since the removal of the Edwards Dam, economists have estimated the total value of Kennebec's recreational fishery at roughly $37,000,000.[53] Recreational fishing on the Penobscot could well exceed that figure. By any measure, the Penobscot River Restoration Project has thus far succeeded in striking a workable balance between river conservation and economic development.

The restoration of such large numbers of alewives has implications for the lower Penobscot ecosystem well beyond the economics of a recreational fishery or the revival of any single species. The elimination of obstructions not only brought back native aquatic species but, more important, also reestablished important ecological connections between saltwater and freshwater ecosystems. Fisheries biologist Richard Dill spoke to the *Bangor Daily News* about this dynamic: "With restoring alewives, there's this huge influx of nutrients coming back from the ocean, connecting the lakes and the rivers back to the ocean . . . So we see this huge shift of energy back to the freshwater ecosystem."[54] In these numbers, alewives support healthy freshwater habitats by cycling nutrients from marine environments back to rivers. Alewives that spawn, perish, and then decompose deposit phosphorus and transfer other nutrients that greatly enrich river habitats.[55] The ecological benefits of strong populations of anadromous fish also extend to terrestrial wildlife. Anadromous fish provide a temporary abundance that is more or less predictable for various predators.[56] At the conclusion of the Kennebec restoration, researchers were encouraged by a resurgence of ospreys and bald eagles along the Kennebec and Sebasticook as a result of the millions of alewives that now inhabit those spaces.[57]

Much of the success of these projects comes from a holistic approach to environmental restoration. Instead of placing the focus on restoring just one celebrated species, fisheries biologists have widened the lens to encompass the ecosystem as a whole. That said, the revival of Maine Atlantic salmon will arguably represent the most challenging component of the project. Maine salmon travel nearly twenty-five hundred miles across the North Atlantic to reach feeding grounds between Labrador and Greenland. Spending the next two years at sea, they are preyed on by species higher on the marine food chain. For this reason, it is important to temper restoration expectations with the recognition that, according to fisheries biologists, "many environmental conditions must fall within suitable ranges for any individual salmon to complete its life cycle." However, perhaps this holistic approach to fisheries conservation may help revive this most celebrated of Penobscot River fish. Scientists theorize that at least eleven different species of anadromous fish coevolved side by side over thousands of years and carved out an ecological niche in the Penobscot partially based on that degree of biodiversity. Alewives, American shad, rainbow smelt, striped bass, blueback herring, sturgeon, and others provide distinct contributions to the food chain and the overall health of the watershed.[58] The smaller, more numerous species act as a buffer against predation for fish struggling to reestablish a presence in the ecosystem. This biodiversity represents a strong indicator of healthy and stable ecosystems. Conversely, as one species begins to disappear, the effects ripple throughout the entire system.

The restoration of Atlantic salmon will require more time and continued stocking in order to pay dividends. After a historically poor year, fisheries biologists were encouraged by the 727 salmon that returned to the river in 2015.[59] In 2019, the Maine Department of Marine Resources counted 1,196 salmon during their seasonal upriver migration.[60] While certainly an improvement, there is still a ways to go in realizing the goal of strong annual salmon runs without the aid of artificial stocking. It can be difficult to envision the Penobscot River brought back to its historic glory, but there is reason for optimism in that these initial successes in alewife and shad restoration could foreshadow a similar outcome for salmon. Together, these more recent conservation programs leave an inspiring thought: rivers have shown an amazing capacity to bounce back from centuries of human interference. Patrick Keliher, the commissioner of the Maine Department of Marine Resources, declared that the Edwards Dam removal "shows directly that while man does his best to knock the snot out of these rivers, they do heal . . . They do come back."[61]

Aldo Leopold famously wrote of the need for a new way of relating to the natural world in *A Sound Country Almanac*. He argued that "a land ethic changes the role of *Homo sapiens* from conqueror of the land-community to plain member and citizen of it."[62] Leopold would be gratified that this ethic has permeated discussions of environmental health and sustainability on the rivers of New England. These restoration projects have been motivated, in some measure, by the desire to reconsider the proper function of rivers in modern society. Do rivers exist for the sole purpose of satisfying economic demands? Do rivers have value beyond the resources they provide to humanity? Can a river operate as a conduit to a new way of thinking that is more sensitive to the value of biodiversity and ecological stability? For the moment, it appears that the second decade of the twenty-first century has produced a distinct shift in thinking about the significance of free-flowing rivers and the fish that inhabit those environments. On the surface, this appears to be a recent development, but these principles were fundamental to a preindustrial river conservation ethic.

For New England, the recent activity surrounding the Kennebec and Penobscot Rivers has revived a historical narrative in which coalitions of local stakeholders and governing authorities worked together to promote healthy riparian ecosystems. This cooperation also happens to mirror nineteenth-century patterns of river management. The construction of large dams on the Merrimack and Kennebec Rivers in the early nineteenth century provoked strong reactions from farmers, fishers, and millers, all of whom feared that wholesale environmental change would negatively impact their lives. It is no surprise that this rural coalition failed to stave off industrialization given its tremendous economic benefits throughout New England. But the feeling that such dramatic tinkering with the natural course of inland waters may cause significant damage never went away. For this reason, historical attention given to rivers—and the disruptions posed by commercial fishing and industrialization—needs to be woven into this modern story of environmental restoration. More than 150 years after the construction of the Edwards Dam, a new coalition of environmental interests addressed the very same debate and built on the work of their nineteenth-century counterparts. It is quite remarkable that these intimate connections to New England's rivers remained alive and well despite centuries of river degradation. Whether they are aware of the region's history or not, those currently invested in river restoration are operating within a context that stretches back to the conservation efforts of preindustrial farmer-fishermen. Just as nineteenth-century coalitions struggled to protect environments that supported rural communities, twenty-first-century conservation programs are seeking to revive those same resources that for generations strengthened human connections to the land and water.

NOTES

Introduction

1. George Perkins Marsh, *Report, Made under the Authority of the Legislature of Vermont, on the Artificial Propagation of Fish*, in *So Great a Vision: The Conservation Writings of George Perkins Marsh*, ed. Stephen C. Trombulak (Hanover, NH: University of New England Press, 2001), 67, 66.

2. Marsh, *Artificial Propagation of Fish*, 66.

3. "An Act to Prevent the Destruction of Fish Called Shad and Alewives, in Mystic River," February 1789, in *Laws Relating to Inland Fisheries in Massachusetts, 1623–1886* (Boston: Wright & Potter Printers, 1887), 29.

4. W. Jeffrey Bolster, *The Mortal Sea: Fishing the Atlantic in the Age of the Sail* (Cambridge, MA: Belknap Press of Harvard University Press, 2012), 2, 48.

5. Matthew McKenzie, *Clearing the Coastline: The Nineteenth-Century Ecological and Cultural Transformations of Cape Cod* (Hanover, NH: University Press of New England, 2011), 40; George Wingate Chase, *History of Haverhill, Massachusetts* (Haverhill, MA: Published by author, 1861), 451; Samuel Orcutt and Ambrose Beardsley, *The History of the Old Town of Derby Connecticut, 1642–1880* (Springfield, MA: Springfield Printing, 1880), 251.

6. John Cumbler, "The Early Making of an Environmental Consciousness: Fish, Fisheries Commissions and the Connecticut River," *Environmental History Review* 15, no. 4 (Winter 1991): 77–78; C. G. Atkins, "The River Fisheries of Maine," in *The Fisheries and Fishery Industries of the United States*, ed. George Brown Goode, sec. 5, vol. 1 (Washington, D.C.: Government Printing Office, 1887), 683, 687.

7. Daniel Vickers, "Competency and Competition: Economic Culture in Early

America," *William and Mary Quarterly* 47, no. 1 (January 1990): 25; Daniel Vickers, "Those Dammed Shad: Would the River Fisheries of New England Have Survived in the Absence of Industrialization?," *William and Mary Quarterly* 61, no. 4 (October 2004): 710.

8. William Cronon, *Changes in the Land: Indians, Colonists, and the Ecology of New England* (New York: Hill and Wang, 1983), 5.

9. Ruth Bogin "Petitioning and the New Moral Economy of Post-Revolutionary America," *William and Mary Quarterly* 45, no. 3 (July 1988): 392.

10. Vickers, "Those Dammed Shad," 688.

11. Carolyn Merchant, *Ecological Revolutions: Nature, Gender, and Science in New England* (Chapel Hill: University of North Carolina Press, 1989), 150–52.

12. Daniel Vickers, *Farmers and Fishermen: Two Centuries of Work in Essex County, Massachusetts, 1630–1850* (Chapel Hill: University of North Carolina Press, 1994), 206.

13. Christopher Clark, *The Roots of Rural Capitalism: Western Massachusetts, 1780–1860* (Ithaca, NY: Cornell University Press, 1990), 15–16.

14. James A. Henretta, "Families and Farms: Mentalité in Pre-Industrial America," *William and Mary Quarterly* 35, no. 1 (January 1978): 19.

15. Vickers, "Those Dammed Shad," 710.

16. Merchant, *Ecological Revolutions*, 150, 153.

17. Richard W. Judd, *Common Lands, Common People: The Origins of Conservation in Northern New England* (Cambridge, MA: Harvard University Press, 1997), 6–7.

18. Brian Donahue, *The Great Meadow: Farmers and the Land in Colonial Concord* (New Haven, CT: Yale University Press, 2004), 22; Brian Donahue, "Environmental Stewardship and Decline in Old New England," *Journal of the Early Republic* 24, no. 2 (Summer 2004): 241.

19. Judd, *Common Lands, Common People*, 5.

20. Bolster, *Mortal Sea*, 56.

21. Gary Kulik, "Dams, Fish, and Farmers: Defense of Public Rights in Eighteenth-Century Rhode Island," in *The Countryside in the Age of Capitalism: Essays in the Social History of Rural America*, ed. Steven Hahn and Jonathan Prude (Chapel Hill: University of North Carolina Press, 1985), 28.

22. Vickers, *Farmers and Fishermen*, 205.

23. John T. Cumbler, *Reasonable Use: The People, the Environment, and the State, New England 1790–1930* (New York: Oxford University Press, 2001), 25.

24. Morton Horwitz, *The Transformation of American Law, 1780–1860* (Cambridge, MA: Harvard University Press, 1977); Theodore Steinberg, *Nature Incorporated: Industrialization and the Waters of New England* (Cambridge: Cambridge University Press, 1991).

25. Garrett Hardin, "The Tragedy of the Commons," *Science* 162, no. 3859 (December 1968): 1244; Elinor Ostrom, "Reformulating the Commons," *Swiss Political Science Review* 6 no. 1 (Spring 2000): 29–30; Elinor Ostrom, Roy Gardner, and James Walker, *Rules, Games, and Common-Pool Resources* (Ann Arbor: University of Michigan Press, 1994), 6.

26. Ostrom, Gardner, and Walker, *Rules, Games*, 7.

27. Elinor Ostrom, *Governing the Commons: The Evolution of Institutions for Collective Action* (Cambridge: Cambridge University Press, 1990), 61; Bonnie J. McCay and Svein Jentoft, "Market or Community Failure? Critical Perspectives on Common Property Research," *Human Organization* 57, no. 1 (1998): 23; James M. Acheson and Jack

Knight, "Distribution Fights, Coordination Games, and Lobster Management," *Comparative Studies in Society and History* 47, no. 1 (January 2000): 209.

28. Ostrom, "Reformulating the Commons," 30.

Chapter One: Channels and Flows

1. John Winthrop, *The Journal of John Winthrop, 1630–1649*, ed. Richard S. Dunn and Laetitia Yeandle (Cambridge, MA: Belknap Press of Harvard University Press, 1996), 168.

2. James T. Kloppenberg, *Toward Democracy: The Struggle for Self-Rule in European and American Thought* (New York: Oxford University Press, 2016), 85–86.

3. John J. Whittlesey, *Law of the Seashore, Tidewaters and Great Ponds in Massachusetts and Maine: Under the Colony Ordinance of 1641–1647* (Boston, 1932), xxxv–xxxvii, xxix–xxxi, Social Law Library, Boston, Massachusetts, KFM2846.W47 1932.

4. Richard W. Judd, *Common Lands, Common People: The Origins of Conservation in Northern New England* (Cambridge, MA: Harvard University Press, 1997), 17–18.

5. Mart R. Gross, Ronald M. Coleman, and Robert M. McDowall, "Aquatic Productivity and the Evolution of Diadromous Fish Migration," *Science* 239, no. 4845 (March 11, 1988): 1291.

6. John Waldman, *Running Silver: Restoring Rivers and Their Great Fish Migrations* (Guilford, CT: Lyons Press, 2013), 12–16.

7. Raymond McFarland, *A History of the New England Fisheries* (New York: D. Appleton, 1911), 219–20.

8. Frederick W. True, "The Alewife Fishery of Cape Cod," in *The Fisheries and Fishery Industries of the United States*, ed. George Brown Goode, sec. 5, vol. 1 (Washington, D.C.: Government Printing Office, 1887), 671.

9. McFarland, *New England Fisheries*, 222.

10. C. G. Atkins, "The River Fisheries of Maine," in Goode, *Fisheries and Fishery Industries*, 687–90; McFarland, *New England Fisheries*, 222.

11. Waldman, *Running Silver*, 22.

12. Rory Saunders, Michael A Hachey, and Clem W. Fay, "Maine's Diadromous Fish Community: Past, Present, and Implications for Atlantic Salmon Recovery," *Fisheries* 31, no. 11 (November 2006): 538; Charles H. Stevenson, *The Shad Fisheries of the Atlantic Coast of the United States* (Washington, D.C.: Government Printing Office, 1899), 106, 110.

13. Marshall McDonald, "The Connecticut and Housatonic Rivers and Minor Tributaries of Long Island Sound," in Goode, *Fisheries and Fishery Industries*, 665.

14. McFarland, *New England Fisheries*, 216.

15. Atkins, "River Fisheries of Maine," 678–79.

16. Atkins, 680–83, 709.

17. Saunders, Hachey, and Fay, "Maine's Diadromous Fish Community," 537.

18. Phillip M. Martson and Myron Gordon, "Notes on Fish and Early Fishing in the Merrimack River System," in *Biological Survey of the Merrimack Watershed*, ed. Earl E. Hoover (Concord, NH: New Hampshire Fish and Game Department, 1938), 187.

19. Atkins, "River Fisheries of Maine," 679.

20. Waldman, *Running Silver*, 36.

Chapter Two: Colonial Encounters

1. James B. Petersen, Brian S. Robinson, Daniel F. Belknap, James Stark, and Lawrence K. Kaplan, "An Archaic and Woodland Period Fish Weir Complex in Central Maine," *Archaeology of Eastern North America* 22 (Fall 1994): 214, 217.

2. Petersen et al., "Archaic and Woodland Period Fish Weir," 217.

3. Thomas Morton, *The New English Canaan*, ed. Charles Francis Adams (Boston: Prince Society, 1883), 232, 222.

4. "An Act to Prevent Nuisance by Hedges, Wears [*sic*] [. . .] Obstructing the Passage of Fish in Rivers," in *Laws Relating to Inland Fisheries in Massachusetts, 1623–1886* (Boston: Wright & Potter, 1887), 7–8.

5. Kathleen J. Bragdon, *Native People of Southern New England, 1500–1650* (Norman: University of Oklahoma Press, 1996), 77.

6. Richard W. Judd, *Second Nature: An Environmental History of New England* (Amherst: University of Massachusetts Press, 2014), 67.

7. Bragdon, *Native People*, 69, 77–78.

8. Lucianne Lavin and Marc Banks, "Connecticut's First Fishermen: LeBeau Fishing Camp and Weir," in *State Archeological Preserve Booklet* (Killingly: Connecticut State Historic Preservation Office, 2008), 4.

9. Morton, *New English Canaan*, 15, 138.

10. Peter A. Thomas, "Contrastive Subsistence Strategies and Land Use as Factors for Understanding Indian-White Relations in New England," *Ethnohistory* 23, no 1 (Winter 1976): 11.

11. William Wood, *New England's Prospect* (Boston: Printed for the Prince Society by John Wilson and Son, 1865), 100.

12. Bragdon, *Native People*, 57, 74.

13. Howard S. Russell, *Indian New England before the Mayflower* (Hanover, NH: University of New England Press, 1980), 14.

14. Shepard Krech, *The Ecological Indian: Myth and History* (New York: W. W. Norton, 1999), 96–97.

15. Deborah McGregor, "Coming Full Circle: Indigenous Knowledge, Environment and Our Future," *American Indian Quarterly* 28, nos. 3 and 4 (Summer/Fall 2004): 389.

16. Michael E. Harkin and David Rich Lewis, introduction to *Native Americans and the Environment: Perspectives on the Ecological Indian*, ed. Michael E. Harkin and David Rich Lewis (Lincoln: University of Nebraska Press, 2007), xx.

17. Mae Noble, Phil Duncan, Darren Perry, Kerry Prosper, Denise Rose, Stephan Schnierer, Gail Tipa, Erica Williams, Rene Woods, and Jamie Pittock, "Culturally Significant Fisheries: Keystones for Management of Freshwater Social-Ecological Systems," *Ecology and Society* 21, no. 2 (June 2016), http://dx.doi.org/10.5751/ES-08353-210222; Sue E. Jackson et al., "'We Like to Listen to Stories about Fish': Integrating Indigenous Ecological and Scientific Knowledge to Inform Environmental Flow Assessments," *Ecology and Society* 19, no. 1 (March 2014), http://dx.doi.org/10.5751/ES-05874-190143.

18. Jeff Bolster, *The Mortal Sea: Fishing the Atlantic in the Age of Sail* (Cambridge, MA: Belknap Press of Harvard University Press, 2012), 13.

19. National Oceanic and Atmospheric Administration, "The History of Salmon," *Final CCC Coho Salmon ESU Recovery Plan*, vol. 1 (Washington, D.C.: U.S. Department of Commerce, September 2012): 15.

20. Diana Muir, *Reflections in Bullough's Pond: Economy and Ecosystem in New England* (Hanover: University of New England Press, 2000), 62.

21. John Smith, *A Description of New England; or, the Observations and Discoveries of Captain John Smith*. (Rochester, NY: Post Express Printing Company, 1898), 6, 25.

22. Smith, *Description of New England*, 9.

23. Thomas Hutchinson, *The History of Massachusetts: From the First Settlement Thereof in 1628 until the Year 1750* (Salem: Thomas C. Cushing, 1828), 407.

24. Jeremy Belknap, *The History of New Hampshire*, vol. 1 (Dover, NH: S. C. Stevens, Ela, and Wadleigh, 1831), 219–20.

25. Bruce J. Bourque, *Twelve Thousand Years: American Indians in Maine* (Lincoln: University of Nebraska Press, 2001), 121, 124.

26. William Cronon, *Changes in the Land: Indians, Colonists, and the Ecology of New England* (New York: Hill and Wang, 1983), 70.

27. Cronon, *Changes in the Land*, 57.

28. Muir, *Reflections in Bullough's Pond*, 62.

29. Thomas, "Contrastive Subsistence Strategies," 14.

30. Belknap, *History of New Hampshire*, 198; Bourque, *Twelve Thousand Years*, 142; Bragdon, *Native People*, 26.

31. Calvin Martin, "The European Impact on the Culture of a Northeastern Algonquian Tribe: An Ecological Interpretation," *William and Mary Quarterly* 31, no. 1 (January 1974): 25.

32. Cronon, *Changes in the Land*, 106.

33. Bolster, *Mortal Sea*, 9.

34. Wood, *New England's Prospect*, 44.

35. Bolster, *Mortal Sea*, 56, 77, 58.

36. "Plymouth Colony Fish Laws," *Laws Relating to Inland Fisheries*, 3–5.

37. Phillip J. Greven Jr., *Four Generations: Population, Land, and Family in Colonial Andover, Massachusetts* (Ithaca, NY: Cornell University Press, 1970), 103.

38. Greven, *Four Generations*, 104, 124.

39. "Act to Prevent Nuisance by Hedges, Wears [*sic*]," 7.

40. "An Act for Rendering More Effectual the Act [. . .] Obstructing the Passage of Fish in River," *Laws Relating to Inland Fisheries*, 8–9.

41. Muir, *Reflections in Bullough's Pond*, 64.

42. "An Act in Addition to Act Made to Prevent the Destruction of the Fish Called Alewives, and Other Fish," *Laws Relating to Inland Fisheries*, 9–10.

43. "An Act in Addition to an Act [. . .] to Prevent the Destruction of the Fish Called Alewives and Other Fish," *Laws Relating to Inland Fisheries*, 11–12.

44. "An Act in Addition to and for Rendering More Effectual the Laws [. . .] Preventing the Destruction of the Fish Called Alewives, and Other Fish," *Laws Relating to Inland Fisheries*, 12–14.

45. "An Act in Addition to Act Made to Prevent the Destruction of the Fish Called Alewives, and Other Fish," 9.

Chapter Three: Farmer and Fisherman

1. Elias Robbins, "Account Book, 1815–1845" (published 1815), in *History of Maine Fisheries*, Paper 13, University of Maine Digital Commons, 12, https://digitalcommons.library.umaine.edu/fisheries/13.

2. James A. Henretta, "Families and Farms: Mentalité in Pre-Industrial America," *William and Mary Quarterly* 35, no. 1 (January 1978): 15.

3. Richard W. Judd, *Second Nature: An Environmental History of New England* (Amherst: University of Massachusetts Press, 2014), 84–85; Carolyn Merchant, *Ecological Revolutions: Nature, Gender, and Science in New England* (Chapel Hill: University of North Carolina Press, 1989), 133.

4. Diana Muir, *Reflections in Bullough's Pond: Economy and Ecosystem in New England* (Hanover, NH: University Press of New England, 2000), 62–63.

5. Brian Donahue, "Environmental Stewardship and Decline in Old New England," *Journal of the Early Republic* 24, no. 2 (Summer 2004): 238.

6. Daniel Vickers, "Competency and Competition: Economic Culture in Early America," *William and Mary Quarterly* 47, no. 1 (January 1990): 4.

7. Seventh Census of the United States, 1850, National Archives Microfilm Publication M432, 1009 rolls, Records of the Bureau of the Census, Record Group 29, National Archives, Washington, D.C., Ancestry.com.

8. Selected Federal Census Non-Population Schedules: Agriculture, 1850, Bowdoinham, Lincoln, Maine, Archive Collection Number: 1–130, roll 1, page 209, line 23, Ancestry.com.

9. Seventh Census of the United States, 1850.

10. Silas Adams, *The History of the Town of Bowdoinham, 1762–1912* (Fairfield, ME: Fairfield Publishing Company, 1912), 217.

11. Brian Donahue, *The Great Meadow: Farmers and the Land in Colonial Concord* (New Haven, CT: Yale University Press, 2004), 197; Merchant, *Ecological Revolutions*, 155.

12. Christopher Clark, "The View from the Farmhouse: Rural Lives in the Early Republic," *Journal of the Early Republic* 24, no. 2 (Summer 2004): 199.

13. Robbins, "Account Book," 75–83.

14. Christopher Clark, *The Roots of Rural Capitalism: Western Massachusetts, 1780–1860* (Ithaca, NY: Cornell University Press, 1990), 29–30.

15. Charles Sellers, *The Market Revolution: Jacksonian American, 1815–1846* (Oxford: Oxford University Press, 1991), 13.

16. Clark, *Roots of Rural Capitalism*, 30.

17. Robbins, "Account Book," 12.

18. Robbins, 19–22.

19. Robbins, 72.

20. Robbins, 72–73, 79–80.

21. Robbins, 121–22.

22. Devereaux Family Papers, "Account Book, 1868–1873" (published 1868), *History of Maine Fisheries*, Paper 72, University of Maine Digital Commons, https://digitalcommons.library.umaine.edu/fisheries/72.

23. Matthew Patten, *The Diary of Matthew Patten of Bedford, N.H., from Seventeen Hundred Fifty-Four to Seventeen Hundred Eighty-Eight* (Concord, NH: Rumford Printing Company, 1903), 1.

24. T. H. Breen, *American Insurgents, American Patriots: The Revolution of the People* (New York: Hill and Wang, 2010), 5, 9.

25. Daniel Vickers, "Those Damned Shad: Would the River Fisheries of New England Have Survived in the Absence of Industrialization?," *William and Mary Quarterly* 61, no. 4 (October 2004): 692.

26. Patten, *Diary*, 110–11.

27. Vickers, "Those Dammed Shad," 693.

28. Sylvester Judd, *History of Hadley, including the Early History of Hatfield, South Hadley, Amherst and Granby, Massachusetts* (Northampton, MA: Printed by Metcalf, 1863), 316.

29. Vickers, "Those Dammed Shad," 694; Patten, *Diary*, 138, 153.

30. Vickers, "Those Dammed Shad," 694; Patten, *Diary*, 285.

31. Vickers, "Those Dammed Shad," 697; Patten, *Diary*, 170–71.

32. Vickers, "Those Dammed Shad," 697.

33. Judd, *Second Nature*, 85.

34. Vickers, "Those Dammed Shad," 710.

35. Patten, *Diary*, 198.

36. Judd, *History of Hadley*, 317.

37. J. W. Meader, *The Merrimack River: Its Source and Its Tributaries* (Boston, MA: B. B. Russell, 1869), 190.

38. Judd, *History of Hadley*, 317.

39. Patten, *Diary*, 303, 301.

40. Harry L. Watson, "'The Common Rights of Mankind': Subsistence, Shad, and Commerce in the Early Republic South," *Journal of American History* 83, no. 1 (June 1996): 40, 32; Christopher Clark, *The Roots of Rural Capitalism: Western Massachusetts, 1780–1860* (Ithaca, NY: Cornell University Press, 1990), 16; Henretta, "Families and Farms," 28.

41. Petition of the Inhabitants of Bradford, May 28, 1802, Bill Packet: An Act in Addition to an Act Entitled "An Act to Regulate the Catching of Salmon [. . .] in the Merrimack River and the Streams Emptying into the Same," February 7, 1803, Massachusetts Archives, Boston, MA (hereafter MassA).

42. "Remonstrance of the Town of Bradford," June 1801, Senate unpassed legislation, 1801 sess., file 2707, MassA.

43. Harold Fisher Wilson, *The Hill Country of Northern New England: Its Social and Economic History* (New York: AMS Press, 1967), 24, 19.

44. Abby Maria Hemenway, *The Vermont Historical Gazetteer: A Magazine Embracing a History of Each Town, Civil, Ecclesiastical, Biographical, and Military*, vol. 2 (Burlington, VT: Published by Miss A. M. Hemenway, 1871), 670.

45. Hemenway, *Vermont Historical Gazetteer*, 2:779.

46. C. Edward Skeen, "The Year without a Summer: A Historical View," *Journal of the Early Republic* 1, no. 1 (Spring 1981): 53, 58, 60.

47. Abby Maria Hemenway, *The Vermont Historical Gazetteer: A Local History of All the Towns in the State, Civil Educational, Biographical, Religious, and Military*, vol. 4 (Burlington, VT: Published by A. M. Hemenway, 1871), 1009.

48. Sarah F. McMahon, "'All Things in Their Proper Season': Seasonal Rhythms of Diet in Nineteenth Century New England," *Agricultural History* 63, no 2. (Spring 1989): 140.

49. Muir, *Reflections in Bullough's Pond*, 62.

50. Hemenway, *The Vermont Historical Gazetteer*, 4:1009–10.
51. McMahon, "'All Thing in Their Proper Season,'" 132.
52. Wilson, *Hill Country*, 22.
53. Henretta, "Families and Farms," 19.
54. Margaret Ellen Newell, "The Birth of New England in the Atlantic Economy: From Its Beginning to 1770," in *Engines of Enterprise: An Economic History of New England*, ed. Peter Temin (Cambridge, MA: Harvard University Press, 2000), 15.
55. Matthew McKenzie, *Clearing the Coastline: The Nineteenth-Century Ecological and Cultural Transformations of Cape Cod* (Hanover, NH: University Press of New England, 2011), 40.
56. Thomas B. Hazard, *Nailer Tom's Diary: The Journal of Thomas B. Hazard of Kingstown Rhode Island, 1778–1840*, ed. Caroline Hazard (Boston: Merrymount Press, 1930), ix.
57. Vickers, "Those Dammed Shad," 698.
58. Hazard, *Nailer Tom's Diary*, xix, xvii.
59. Hazard, xviii.
60. Hazard, 504–6.
61. Hazard, 507.
62. Christopher Clark, "Rural America and the Transition to Capitalism," *Journal of the Early Republic* 16, no. 2 (Summer 1996): 226.
63. Vickers, "Those Dammed Shad," 695.
64. John J. McCusker and Russell R. Menard, *The Economy of British North America, 1607–1789* (Chapel Hill: University of North Carolina Press, 1991), 145.
65. Henry R. Stiles, *The History of Ancient Wethersfield* (New York: Grafton Press, 1904), 535, 541.
66. Stiles, *History of Ancient Wethersfield*, 498.
67. Justus Riley, Schooner Polly Voyage to Cape Francois, June 1792, Maritime Collection, 1639–1987, Series II: History & People, Wethersfield Historical Society, Wethersfield, Connecticut (hereafter cited as WHS).
68. Justus Riley, Account of Salmon Bought for Justus Riley and Company, April 12, 1783, Maritime Collection, 1639–1986, Series III: Vessels: Sloop Nancy, WHS; Justus Riley, Account of Sundry Bill in Salmon Voyage in Sloop Nancy, June 8, 1783, Maritime Collection, 1639–1986, Series III: Vessels: Sloop Nancy, WHS.
69. Accounts of Salmon on Board Sloop Black Joke, April 19, 1783, Maritime Collection, 1639–1986, Series III: Vessel Black Joke, WHS; Accounts of Salmon on Board Sloop Dolphin, May 13, 1783, Maritime Collection, 1639–1986, Series III: Vessel Dolphin, WHS.
70. Accounts of Salmon on Board Sloop Black Joke, April 19, 1783, WHS.
71. An Annual Return of the Number of Barrels Pickled Fish Inspected in the Commonwealth of Massachusetts, April 1804 to June 1805, Returns from Public Inspectors, SC1/Series 139X, MassA.
72. An Annual Return of the Number of Casks of Pickled Fish Inspected in the Commonwealth of Massachusetts, June 1812 to January 1813, Returns from Public Inspectors, SC1/Series 139X, MassA.
73. An Annual Return of the Number of Casks of Pickled Fish Inspected in the Commonwealth of Massachusetts," May 1823 to April 1824, Returns from Public Inspectors, SC1/Series 139X, MassA.
74. Letter from Charles Buck to Edward Russell, January 26, 1830, Fish Returns, box 4, Hampden Maine, 1828, Maine State Archives, Augusta, ME.

75. C. G. Atkins, "The River Fisheries of Maine," in *The Fisheries and Fish Industries of the United States*, ed. George Brown Goode, sec. 5, vol. 1 (Washington. D.C.: Government Printing Office, 1887), 683, 687.

76. John Cumbler, "The Making of an Environmental Consciousness: Fish, Fisheries Commissions and the Connecticut River," *Environmental History Review* 15, no. 4 (Winter 1991): 77.

77. Petition of Inhabitants of Lancaster, Dartmouth, Northumberland, and Stratford, Fishing in the Connecticut River, May 17, 1788, New Hampshire State Archives, Concord, NH.

78. Richard W. Judd, *Common Lands, Common People: The Origins of Conservation in Northern New England* (Cambridge, MA: Harvard University Press, 1997), 62–63.

Chapter Four: Complications in the Commons

1. Petition of James Cochran and Others, August 1811, Bill Packet: Act in Further Addition to an Act Entitled "An Act to Regulate the Catching [. . .] Merrimack River," February 1812, Massachusetts Archives, Boston, MA (hereafter MassA).

2. "An Act in Further Addition [. . .] to Regulate the Catching of Salmon, Shad, and Alewives [. . .] Merrimack River," February 29, 1812, in *A Collection of the Laws of Massachusetts Relating to Inland Fisheries, 1623–1886* (Boston: State Printers, 1887), 130–31.

3. "An Act More Effectually to Prevent the Destruction of Alewives [. . .] in the Towns of Salem and Danvers," in *A Collection of the Laws of Massachusetts Relating to Inland Fisheries, 1623–1886* (Boston: State Printers, 1887), 9, 14.

4. Petition of James Cochran.

5. John T. Cumbler, *Reasonable Use: The People, the Environment, and the State, New England 1790–1930* (New York: Oxford University Press, 2001), 78.

6. "An Act to Regulate the Catching of Salmon [. . .] Merrimack River," October 1783, in *Acts and Resolves of the Commonwealth of Massachusetts* (Boston: Wright & Potter, 1896), 547.

7. Cumbler, *Reasonable Use*, 78.

8. Petition of Inhabitants of This State Living on or near the River Merrimack, Regulations for Taking of Fish," June 1790, New Hampshire State Archives, Concord, NH (hereafter NHA).

9. Ruth Bogin, "Petitioning and the New Moral Economy of Post-Revolutionary America," *William and Mary Quarterly* 45, no. 3 (July 1988): 392.

10. Alan Taylor, *Liberty Men and Great Proprietors: The Revolutionary Settlement on the Maine Frontier, 1760–1820* (Chapel Hill: University of North Carolina Press, 1990), 96.

11. "Petition of Inhabitants of Bradford," June 1812, Senate unpassed legislation, 1813 sess., file 4778, MassA.

12. Remonstrance of the Inhabitants of Chelmsford, June 8, 1812, Senate unpassed legislation, 1813 sess., file 4778, MassA.

13. E. P. Thompson, "The Moral Economy of the English Crowd in the Eighteenth Century," *Past and Present*, no. 50 (1971): 79.

14. Thompson, "Moral Economy," 78.

15. "Catching of salmon [. . .] Merrimack River," 546–47; "An Act to Prevent the Destruction and to Regulate . . . Salmon, Shad, and Alewives," February 1789,

in *Acts and Resolves of the Commonwealth of Massachusetts* (Boston: Wright & Potter, 1894), 157.

16. "An Act for Reviving an Act [. . .]," May 11, 1767, in *Laws of New Hampshire*, vol. 3, *Province Period, 1745–1774* (Bristol, NH: Musgrove Printing House, 1915), 407.

17. Petition of the Undersigned Inhabitants of This State, June 1797, NHA.

18. "An Act in Addition to an Act, Entitled 'An Act to Prevent the Destruction of Salmon, Shad, and Alewives in Merrimack River' [. . .]," December 20, 1797, in *Laws of New Hampshire*, vol. 6, *Second Constitutional Period, 1792–1801* (Concord, NH: Evans Printing, 1917), 476–77.

19. "Petition of the Undersigned Inhabitants."

20. Remonstrance of the Town of Bradford, June 1801, Senate unpassed legislation, 1801 sess., file 2707, MassA.

21. "An Act to Promote the Increase of the Fish Call'd Alewives in Great Cohass Brook in Derryfield," March 23, 1776, in *Laws of New Hampshire*, vol. 4, *Revolutionary Period, 1776–1784* (Bristol, NH: Musgrove Printing, 1916), 8.

22. "An Act to Regulate the Catching of Salmon [. . .] Merrimack River," June 1783, in *Acts and Resolves of the Commonwealth of Massachusetts, 1782–1783* (Boston: Wright & Potter, 1894), 546.

23. "Catching of Salmon [. . .] Merrimack River," 547–49.

24. "An Act to Prevent the Destruction of Salmon, Shad, and Alewives in the Merrimack River," April 9, 1784, in *Laws of New Hampshire*, 4:547.

25. "An Act in Amendment of, and in Addition to an Act, Intitled [*sic*] 'An Act to Prevent the Destruction of Salmon, Shad, & Alewives in the Merrimack River [. . .],'" June 26, 1786, in *Laws of New Hampshire*, vol. 5, *First Constitutional Period, 1784–1792* (Concord, NH: Rumford Press, 1916), 186–87.

26. "An Act to Regulate the Catching of Salmon, Shad, and Alewives [. . .] Merrimack River," March 4, 1790, in *Acts and Resolves of the Commonwealth of Massachusetts, 1788–1789* (Boston: Wright &Potter, 1894), 499–500.

27. "An Act to Prevent the Destruction of Salmon, Shad, and Alewives in Merrimack River and for Repealing All the Laws Heretofore Made for that Purpose," June 18, 1790, in *Laws of New Hampshire*, 5:527–29; "An Act in Addition to, and in Amendment of an Act, Intitled 'An Act to Prevent the Destruction [. . .] in Merrimac [*sic*] River [. . .],'" January 12, 1795, in *Laws of New Hampshire*, 5:221–23.

28. "Petition of the Undersigned Inhabitants."

29. Elinor Ostrom, "Reformulating the Commons," *Swiss Political Science Review* 6, no.1 (Spring 200): 30.

30. Amendments and revisions to Merrimack River statutes appear in legislative record for the years 1803, 1804, 1811, 1812, 1817, 1819, 1832, and 1834, in *A Collection of the Laws of Massachusetts Relating to Inland Fisheries, 1623–1886* (Boston: State Printers, 1887).

31. C. G. Atkins, "The River Fisheries of Maine," in *The Fisheries and Fish Industries of the United States*, ed. George Brown Goode, sec. 5, vol. 1 (Washington, D.C.: Government Printing Office, 1887), 727.

32. Sylvester Judd, *The History of Hadley: Including the Early History of Hatfield, South Hadley, Amherst, and Granby* (Northampton, MA: Printed by Metcalf, 1863), 316.

33. Atkins, "River Fisheries of Maine," 727.

34. Petition of Inhabitants of Lancaster, Dartmouth, Northumberland, and Stratford, Fishing in the Connecticut River, May 17, 1788, NHA.

35. Justus Riley, Account of Salmon Bought for Justus Riley and Company, April 12, 1783, Maritime Collection, 1639–1986, Series III: Vessels: Sloop Nancy, Wethersfield Historical Society.

36. Petition of Inhabitants of Lancaster, Dartmouth, Northumberland, and Stratford.

37. Theodore Lyman and Alfred A. Reed, Report to the Senate concerning the Obstructions to the Passage of Fish in the Connecticut and Merrimack Rivers, S.Doc. No. 8 (Boston, 1866), Massachusetts Historical Society, Boston, MA, 32–33.

38. Petition of Inhabitants of Lancaster, Dartmouth, Northumberland, and Stratford.

39. Lyman and Reed, "Report to the Senate," 32.

40. Petition of Inhabitants of Lancaster, Dartmouth, Northumberland, and Stratford.

41. "An Act to Prevent the Destruction of Salmon and Shad in Connecticut River," February 6, 1789, in *Laws of New Hampshire*, 5:402–4.

42. Subsequent amendments and revisions governing New Hampshire's Connecticut River fisheries appeared in 1791 and 1795.

43. "An Act to Prevent the Destruction of Salmon and Shad in Connecticut River," June 20, 1788, in *Acts and Laws of the Commonwealth of Massachusetts* (Boston: Wright & Potter, 1894), 28–30.

44. "Destruction of Salmon and Shad," 28–30.

45. "Resolve Suspending the Law Regulating the Fishery in Connecticut River [. . .]," June 9, 1789, in *Acts and Laws of the Commonwealth of Massachusetts* (Boston: Wright & Potter, 1894), 543.

46. "Resolve Suspending the Law Regulating the Fishery in Connecticut River and Requesting the Governor to Write to the Governor of Connecticut Relative Thereto," June 1, 1790, in *Acts and Laws of the Commonwealth of Massachusetts* (Boston: Wright & Potter, 1895), 96.

47. "Resolve Suspending the Law for Regulating the Fishery in Connecticut River, June 1, 1790, 96.

48. "An Act for Regulating the Fishery in Connecticut River," March 7, 1791, in *Acts and Laws of the Commonwealth of Massachusetts* (Boston: Wright & Potter, 1895), 58.

49. "Resolve Requesting [. . .] Write to New Hampshire and Vermont for Preservation of Fish in CT River," June 1791, in *Acts and Laws of the Commonwealth of Massachusetts* (Boston: Wright & Potter, 1895), 58.

50. "An Act for Regulating the Fishery in Connecticut River," December 10, 1791, in *Laws of New Hampshire*, 5:788–89.

51. "An Act Regulating Fisheries," March 1787, in *Statutes of the State of Vermont* (Bennington, VT: Printed by Anthony Haswell, 1791), 80–81.

52. Petition of Residents of Bristol, Vermont. Use of Seines, Net, Scoop Net, September 17, 1832, Vermont State Papers, vol. 62, p. 88, Vermont State Archives, Middlesex, VT (hereafter VTA); Petition of Residents of Ludlow, Vermont. Proposal for Regulatory Law," September 4, 1832, Vermont State Papers, vol. 62, p. 85, VTA.

53. "An Act to Preserve Fish . . . Starksborough," October 28, 1829, in *Acts Passed by the Legislature of the State of Vermont* (Woodstock, VT: D. Watson Printer, 1829), 152.

54. J. W. Meader, *The Merrimack River: Its Source and Its Tributaries* (Boston, MA: B. B. Russell, 1869), 248–49.

55. "An Act to Regulate the Catching of Salmon," March 4, 1790, in *Acts and Laws of the Commonwealth of Massachusetts, 1788–1789* (Boston: Wright and Potter Printing Company, 1894), 498–503.

56. Cumbler, *Reasonable Use*, 147–48.
57. Burnham v. Webster, 5 Mass. 266 (1809).
58. Petition of James Cochran.
59. "An Act in Further Addition to an Act . . . Merrimack River," February 29, 1812, in *Acts and Resolves of the Commonwealth of Massachusetts* (Boston: Printed by Adams, Rhoades, 1912), 614–15.
60. Richard Judd, *Second Nature: An Environmental History of New England* (Amherst, MA: University of Massachusetts Press, 2014), 185.
61. "An Act Regulating the Taking the Fish Called Alewives [. . .] Haverhill," February 9, 1803, in *Acts and Laws of the Commonwealth of Massachusetts* (Boston: Wright & Potter, 1898), 94.
62. "Taking the Fish Called Alewives," 94.
63. Second Census of the United States (Washington, D.C: Duane Printer, 1801), 8.
64. Raymond McFarlane, *A History of the New England Fisheries* (New York: J. F. Tapley, 1911), 198.
65. McKenzie, *Clearing the Coastline: The Nineteenth-Century Ecological and Cultural Transformation of Cape Cod* (Hanover, NH: University of New England Press, 2011), 38–39.
66. Petition of the Inhabitants of Bradford, May, 28, 1802, Bill Packet: An Act in Addition to an Act Entitled "An Act to Regulate the Catching of Salmon . . . in the Merrimack River and the Streams Emptying into the Same," MassA.
67. "An Act in Addition to an Act, Entitled 'An Act to Regulate the Catching of Salmon, Shad, and Alewives in Merrimack River [. . .]," February 7, 1803, in *Acts and Laws of the Commonwealth of Massachusetts* (Boston: Wright & Potter, 1898), 81–82; "An Act to Regulate the Taking of Fish Called Alewives in Johnston's Brook So Called, Emptying into the Merrimack River, in the Town of Bradford [. . .]," February 12, 1803, in *Acts and Laws of the Commonwealth of Massachusetts* (Boston: Wright & Potter, 1898), 117.
68. Petition of Inhabitants of Chester, bounding by Massabesick Pond, November 1800, NHA.
69. "An Act for the Preservation of Alewives in Cocheco River," June 27, 1816, in *Laws of New Hampshire*, vol. 8, *Second Constitutional Period, 1811–1820* (Concord, NH: Evans Printing, 1920), 513; "An Act for the Preservation of Alewives in Salmon-Fall River," June 25, 1818, in *Laws of New Hampshire*, 8:706.
70. Jamie H. Eves, "'Shrunk to a Comparative Rivulet': Deforestation, Stream Flow, and Rural Milling in 19th-Century Maine," *Technology and Culture* 33, no. 1 (January 1992): 43.
71. Gary Kulik, "Dams, Fish, and Farmers: Defense of Public Rights in Eighteenth-Century Rhode Island," in *The Countryside in the Age of Capitalist Transformation: Essays in the Social History of Rural America*, ed. Steven Hahn and Jonathan Prude, 25–50 (Chapel Hill: University of North Carolina Press, 1985), 36.
72. "An Act Impowering Sundry Committees to Cause Sluices to Be Made, in All Dams, Made Across Such Part of Beaver Brook as Is in the Colony of New Hampshire [. . .]," July 3, 1776, in *Laws of New Hampshire*, 4:20.
73. "An Act Impowering Sundry Committees to Cause Sluices to be Made."
74. "Petition of Winchester, Hinsdale, and Swanzey, Fishing in the Ashuelot River," May 30, 1788, NHA.
75. "New Hampshire House of Representatives, Committee Report: Dam on Ashuelot River," June 12, 1788, NHA.

76. "An Act for Opening Sluiced in Each Dam—Across the Ashuelet River So That Salmon and Other Fish May Have Free Passage through the Same from Connecticut River," in *Laws of New Hampshire*, 5:343.

77. Petition of the Selectmen and Justices of the Peace of the Towns of Gilmanton, Preservation of Fish in the Merrimack River, June 1790, NHA.

78. Petition of the Selectmen and Justices [. . .] of Gilmanton.

79. Petition of Samuel Ladd of Gilmanton and Others of Sanbornton, Regulation of Mills," June 4, 1790, NHA.

80. Petition of William Adams Esq. and Residents of Middlesex County, January 1804, Bill Packet: An Act in Addition to an Act Entitled "An Act to Regulate the Catching . . . Merrimack River," March 9, 1804, MassA.

81. "Petition of William Adams Esq."

82. Remonstrance of the Fish Wardens of Chelmsford and Westford, January 1804, Bill Packet: An Act in Addition to an Act entitled "An Act to regulate the catching . . . Merrimack River," March 9, 1804, MassA.

83. Report of a Committee of Both Houses, February 1804, Bill Packet, An Act in Addition to an Act entitled "An Act to regulate the catching . . . Merrimack River," March 9, 1804, MassA.

84. Committee of Both Houses Report on Stoney Brook Mill Dam, March 1804, Bill Packet: An Act in Addition to an Act Entitled "An Act to Regulate the Catching . . . Merrimack River," March 9, 1804, MassA.

85. Kulik, "Dams, Fish, and Farmers," 33.

86. "Petition of Inhabitants of Derryfield and Adjacent Town, Fishing in Cohass Brook," June 7, 1800, NHA.

87. Diana Muir, *Reflections in Bullough's Pond: Economy and Ecosystem in New England* (Hanover, NH: University of New England Press, 2000), 62.

88. Theodore L. Steinberg, "Dam-Breaking in the 19th-Century Merrimack Valley: Water, Social Conflict, and the Waltham-Lowell Mills," *Journal of Social History* 24, no. 1 (Autumn 1990): 27–28.

89. "Catching of Salmon [. . .] Merrimack River," June 1783, 546; Petition of the Undersigned Inhabitants; An Act to Revive an Act Passed [. . .] Entitled "An Act to Promote the Increase of Fish called Alewives in Great Cohass Brook in Derryfield," April 6, 1781, in *Laws of New Hampshire*, 4:386.

Chapter Five: "From time immemorial"

1. Henry David Thoreau, *A Week on the Concord and Merrimack Rivers* (Boston: James R. Osgood, 1873), 98.

2. Diana Muir, *Reflections in Bullough's Pond: Economy and Ecosystem in New England* (Hanover, NH: University of New England Press), 112.

3. Muir, *Reflections in Bullough's Pond*, 112.

4. Thoreau, *Week on the Concord*, 3; Brian Donahue, "'Dammed at Both Ends and Cursed in the Middle': The 'Flowage' of the Concord River Meadows, 1798–1862," *Environmental Review* 13, no. 3/4 (Autumn/Winter 1989): 52.

5. Thoreau, *Week on the Concord*, 44; Donahue, "Dammed at Both Ends," 58.

6. Thoreau, *Week on the Concord*, 38.

7. Theodore Steinberg, *Nature Incorporated: Industrialization and the Waters of New England* (Cambridge: Cambridge University Press, 1991), 45.

8. Lance E. Davis and H. Louis Stettler, "The New England Textile Industry, 1825–1860: Trends and Fluctuations," in *Output, Employment, and Productivity in the United States after 1800*, ed. Dorothy Brady (Cambridge, MA: National Bureau of Economic Research, 1966), 229.

9. Morton J. Horwitz, *The Transformation of American Law, 1780–1860* (Cambridge, MA: Harvard University Press, 1977), 47.

10. Steinberg, *Nature Incorporated*, 79.

11. John Cumbler, *Reasonable Use: The People, the Environment, and the State, New England 1790–1930* (New York: Oxford University Press, 2001), 66.

12. Horwitz, *Transformation of American Law*, 34; Gary Kulik, "Dams, Fish, and Farmers: Defense of Public Rights in Eighteenth-Century Rhode Island," in *The Countryside in the Age of Capitalist Transformation: Essays in the Social History of Rural America*, ed. Steven Hahn and Jonathan Prude (Chapel Hill: University of North Carolina Press, 1985), 26.

13. Steinberg, *Nature Incorporated*, 65, 79.

14. New Hampshire Department of Environmental Services, "Environmental Fact Sheet: The Lower Merrimack River" (Concord, NH, 2008).

15. Steinberg, *Nature Incorporated*, 56–57; Petition from Dudley A. Tyng and the Proprietors of Locks and Canals on the Merrimack River, Senate unpassed legislation, 1801 sess., file 2707, Massachusetts Archives, Boston, MA (hereafter MassA); James Grant Wilson and John Fiske, *Appleton's Cyclopedia of American Biography*, vol. 6 (New York: D. Appleton & Sons, 1889), 202.

16. Remonstrance of the Town of Dracut, June 1801, Senate unpassed legislation, 1801 sess., file 2707, MassA.

17. Second Census of the United States (Washington, D.C.: Duane Printer, 1801), 8–9.

18. A. Howard Clark, "The Fisheries of Massachusetts," in *The Fisheries and Fish Industries of the United States*, sec. 2, *A Geographical Review of the Fisheries Industries and Fishing Communities for the Year 1880*, ed. George Brown Goode (Washington, D.C.: Government Printing Office, 1887), 132.

19. Remonstrance of the Town of Bradford, June 1801, Senate unpassed legislation, 1801 sess., file 2707, MassA.

20. Remonstrance of the Town of Bradford.

21. Remonstrance of the Town of Andover," June 1801, Senate unpassed legislation, 1801 sess., file 2707, MassA.

22. Remonstrance of the Town of Chelmsford," June 1801, Senate unpassed legislation, 1801 sess., file 2707, MassA.

23. Remonstrance of the Town of Chelmsford.

24. Remonstrance of the Town of Dracut.

25. Petition from John Ford and the Proprietors of the Middlesex and Pawtucket Canals," Senate unpassed legislation, 1813 sess., file 4778, MassA.

26. Remonstrance of the Inhabitants of Chelmsford," June 8, 1812, Senate unpassed legislation, 1813 sess., file 4778, MassA.

27. Remonstrance of the inhabitants of Chelmsford, 1812.

28. Remonstrance of Inhabitants of Bradford, June 1812, Senate unpassed legislation, 1813 sess., file 4778, MassA.

29. Remonstrance of William Adams Esq., June 1812, Senate unpassed legislation, 1813 sess., file 4778, MassA.

30. Remonstrance of the Counties of Rockingham and Hillsborough, New Hampshire, May 27, 1812, Senate unpassed legislation, 1813 sess., file 4778, MassA.

31. Petition of Elijah Johnston, Weirs and Penobscot River, June 17, 1820, box 6, folder 20, Maine State Archives, Augusta, ME (hereafter MSA).

32. Petition of the Chiefs and Others of the Penobscot Indian Tribe, Weirs and the Penobscot River," January 24, 1821, box 8, folder 16, MSA.

33. C. G. Atkins, "The River Fisheries of Maine," in *The Fisheries and Fish Industries of the United States*, ed. George Brown Goode, sec. 5, vol. 1 (Washington, D.C.: Government Printing Office, 1887), 680–81.

34. Atkins, "River Fisheries of Maine," 683.

35. George Brown Goode, U.S. Commission of Fish and Fisheries, *Report of the Commissioner for 1872 and 1873* (Washington, D.C.: Government Printing Office, 1874), 964.

36. An Act to Prevent the Destruction of the Fish Called Salmon, Shad, and Alewives in Kennebec River [. . .], July 1786, in *Acts and Laws of the Commonwealth of Massachusetts* (Boston: Wright & Potter, 1893), 57.

37. An Act to Prevent the Destruction and to Regulate the Catching of the Fish Called Salmon, Shad, and Alewives in the Rivers and Streams in the Counties of Cumberland and Lincoln [. . .], February 16, 1789, in *Acts and Laws of the Commonwealth of Massachusetts* (Boston: Wright & Potter, 1894), 157.

38. "An Act for the Preservation of Salmon, Shad, and Alewives, in Penobscot River, and the Streams Emptying into Said River [. . .]," February 26, 1810, in *Acts and Resolves of the Commonwealth of Massachusetts* (Boston: State Printers, 1886), 329–30.

39. "An Act for the Preservation of Fish in Penobscot River and Bay, and the Several Streams Emptying into the Same," February 22, 1814, in *Laws of the Commonwealth of Massachusetts* (Boston: State Printers, 1890), 398.

40. Petition of the Town of Frankfort, Weirs in the Penobscot River, February 4, 1821, box 8, folder 16, MSA.

41. Remonstrance of Weir Fishermen of Bucksport and Frankfort, January 22, 1821, box 8, folder 19, MSA.

42. Petition of the Town of Georgetown, Weirs and the Kennebec River, February 7, 1821, box 8, folder 15, MSA.

43. Petition of the Town of Phippsburg, Weirs at the Mouth of the Kennebec River," February 6, 1821, box 8, folder 15, MSA.

44. Petition of Greenleaf White and Others, Requesting an Act of Incorporation for the Purposes of Erecting a Dam across the Kennebec River, January 9, 1834, box 107, chapter 134, MSA.

45. Petition of Ebenezer Bacon and Others, Residents of Waterville against the Dam at Augusta, February 4, 1834, box 107, chapter 134, MSA.

46. Remonstrance of Joseph Look et al., Waterville against Greenleaf White's Dam Proposal, February 17, 1834, box 107, chapter 134, MSA.

47. Remonstrance of Arthur Barry and 306 Others, against Dam at Augusta, February 12, 1834, box 107, chapter 134, MSA.

48. Remonstrance of the Town of Georgetown, against the Petition of Greenleaf White for a Dam across the Kennebec River, February 11, 1834, box 107, chapter 134, MSA.

49. Remonstrance of Arthur Barry.

50. Remonstrance of the town of Georgetown.

51. Remonstrance of the Town of Woolwich, against a Dam across the Kennebec, January 30, 1834, box 107, chapter 134, MSA.

52. Remonstrance of the Town of Georgetown.

53. Edward P. Ames and John Lichter, "Gadids and Alewives: Structure within Complexity in the Gulf of Maine," *Fisheries Research* 141 (2013): 77.

54. Remonstrance of the Town of Woolwich.

55. Remonstrance of the Town of Woolwich.

56. Report of Joint Special Committee, Recommending Petition of Greenleaf White," February 20, 1834, box 107, chapter 134, MSA.

57. John S. C. Cabot, *The History of Maine* (Augusta: Brown Thurston, 1892), 412.

58. "An Act to Incorporate the Merrimack Manufacturing Company," February 6, 1822, in *Laws of the Commonwealth of Massachusetts* (Boston: State Printers, 1822), 631.

59. Steinberg, *Nature Incorporated*, 88.

60. Thomas Dublin, *Women at Work: The Transformation of Work and Community in Lowell, Massachusetts, 1826–1860* (New York: Columbia University Press, 1979), 19.

61. Petition of John Tenney and Others, May 1833, Bill Packet: Act to Repeal All Laws Heretofore Made for Regulating Alewife Fishery . . . Dracut and Methuen, March 25, 1834, MassA.

62. Petitions of the Towns of Benjamin Brown and Others, and Petition of Thomas H. Batch and Others," January 1846, Bill Packet: "Act to Repeal the Laws Regulating the Fishery in Merrimack River," April 7, 1846, MassA.

63. Thoreau, *Week on the Concord*, 35.

64. Theodore Lyman and Alfred A. Reed, "Report to the Senate concerning Obstructions to the Passage of Fish in the Connecticut and Merrimack Rivers," S.Doc. No. 8 (Boston, 1866), Massachusetts Historical Society, Boston, MA, 36.

Chapter Six: Rivers Restored

1. John Cumbler, *Reasonable Use: The People, the Environment, and the State, New England, 1790–1930* (Oxford: Oxford University Press, 2001), 74–75.

2. George Perkins Marsh, *Man and Nature: or, Physical Geography as Modified by Human Action*, in *So Great a Vision: The Conservation Writings of George Perkins Marsh*, ed. Stephen C. Trombulak (Hanover, NH: University of New England Press, 2001), 138.

3. George Perkins Marsh, "Report, Made under the Authority of the Legislature of Vermont, on the Artificial Propagation of Fish," in Trombulak, *So Great a Vision*, 66.

4. David Lowenthal, *George Perkins Marsh: Prophet of Conservation* (Seattle: University of Washington Press, 2000), 183.

5. Lowenthal, *George Perkins Marsh*, 183–84.

6. Marsh, "Report," 68, 69.

7. Richard Judd, *Common Lands, Common People: The Origins of Conservation in Northern New England* (Cambridge, MA: Harvard University Press, 1997), 151–54.

8. Cumbler, *Reasonable Use*, 79–80.

9. Theodore Lyman and Alfred A. Reed, *Report to the Senate concerning the Obstructions to the Passage of Fish in the Connecticut and Merrimack Rivers*, S. Doc. No. 8 (Boston, 1866), Massachusetts Historical Society, Boston, MA, 10.

10. Lyman and Reed, *Report to the Senate*, 9.

11. Cumbler, *Reasonable Use*, 93.

12. Cumbler, 93.

13. Lawrence Stolte, *The Forgotten Salmon of the Merrimack* (Department of the Interior, Northeast Region, 1981), 21, 55.

14. Bill Trotter, "Federal Officials Mark 125 Years of Breeding Salmon in Orland," *Bangor (ME) Daily News*, November 22, 2014.

15. Stolte, *Forgotten Salmon*, 41.

16. Judd, *Common Lands, Common People*, 163.

17. Technical Committee for Anadromous Fishery Management of the Merrimack River Basin, "Anadromous Fish Restoration Program: Merrimack River," *Strategic Plan and Status Review*, October 16, 1977, 22.

18. "Anadromous Fish Restoration Program," 19.

19. Judd, *Common Lands, Common People*, 163–64.

20. Cumbler, *Reasonable Use*, 98–99.

21. "Anadromous Fish Restoration Program," 20.

22. Lyman and Reed, *Report to the Senate*, 28; Judd, *Common Lands, Common People*, 163.

23. Judd, *Common Lands, Common People*, 207–8.

24. Judd, 198.

25. Cumbler, *Reasonable Use*, 54, 91, 119.

26. Samuel P. Hays, *Beauty, Health, and Permanence: Environmental Politics in the United States, 1955–1985* (New York: Cambridge University Press, 1987), 3.

27. Jeff Crane, "'Setting the River Free': The Removal of the Edwards Dam and the Restoration of the Kennebec River," *Water History* 1, no. 2 (2009): 132.

28. Howard L. Jelks et al., "Conservation Status of Imperiled North American Freshwater and Diadromous Fishes," *Fisheries* 33, no. 8 (August 2008): 373.

29. Karen E. Limburg and John R. Waldman, "Dramatic Declines in North Atlantic Diadromous Fishes," *Bioscience* 59, no. 11 (December 2009): 959.

30. Rory Saunders, Michael A. Hachey, and Clem W. Fay, "Maine's Diadromous Fish Community: Past, Present, and Implications for Atlantic Salmon Recovery," *Fisheries* 31, no. 11 (November 2006): 541–42.

31. Limburg and Waldman, "Dramatic Declines," 960–61.

32. Dave Owen and Colin Apse, "Trading Dams," *UC Davis Law Review* 48, no. 3 (February 2015): 1053.

33. Yvon Chouinard, "Tear Down 'Deadbeat' Dams," *New York Times*, May 7, 2014.

34. Limburg and Waldman, "Dramatic Declines," 962.

35. John Waldman, *Running Silver: Restoring Rivers and Their Great Fish Migrations* (Guilford, CT: Lyons Press, 2013), 5.

36. Oswald J. Schmitz, *Ecology and Ecosystem Conservation* (Washington, D.C.: Island Press, 2007), 109.

37. Crane, "Setting the River Free," 133–35.

38. Keith Edwards, "Dam's Removal in Augusta Is a Wildlife Success Story," *Portland (ME) Press Herald*, July 1, 2014.

39. J. E. O'Connor, J. J. Duda, and G. E. Grant, "1000 Dams Down and Counting: Dam Removals Are Reconnecting Rivers in the United States," *Science*, May 1, 2015, 496.

40. Michelle Nijhuis, "World's Largest Dam Removal Unleashes U.S. River after Century of Energy Production," *National Geographic*, August 27, 2014.

41. Jeff Crane, *Finding the River: An Environmental History of the Elwha* (Corvallis, OR: Oregon State University Press, 2011), 138–39.

42. Limburg and Waldman, "Dramatic Decline," 963.

43. National Resources Council of Maine and Penobscot River Restoration Trust, "Penobscot River Restoration Trust Fact Sheet," https://conservationgateway.org/Files/Pages/penobscot-river-restorati.aspx.

44. Kevin Miller, "Two Years after Dam's Removal, Penobscot River Flourishes," *Portland (ME) Press Herald*, September 17, 2015.

45. Owen and Apse, "Trading Dams," 1076.

46. Murray Carpenter, "Dam Removal to Help Restore Spawning," *New York Times*, June 11, 2012.

47. Owen and Apse, "Trading Dams," 1076–78.

48. Nick McCrea, "Crews Begin Removing Great Works Dam," *Bangor (ME) Daily News*, June 11, 2012.

49. John Holyoke, "Breaching of Veazie Dam Begins as Part of Penobscot River Restoration," *Bangor (ME) Daily News*, July 22, 2013.

50. Frank Speck, *Penobscot Man: The Life History of a Forest Tribe in Maine* (Philadelphia: University of Pennsylvania Press, 1940), 85.

51. National Resources Council of Maine and Penobscot River Restoration Trust, "Fact Sheet."

52. Miller, "Two Years after Dam's Removal."

53. Jesse Lance Robbins and Lynne Y. Lewis, "Demolish It and They Will Come: Estimating the Economic Impacts of Restoring a Recreational Fishery," *Journal of the American Water Resources Association* 44, no. 6 (December 2008): 1494.

54. John Holyoke, "Alewives Return to Blackman Stream, Inspire Celebratory Festival," *Bangor (ME) Daily News*, May 22, 2015.

55. Waldman, *Running Silver*, 16.

56. Mary F. Wilson and Karl C. Halupka, "Anadromous Fish as Keystone Species in Vertebrate Communities," *Conservation Biology* 9, no. 3 (June 1995): 494.

57. John Holyoke, "Edwards Dam Success Foreshadows Penobscot River Project's Future," *Bangor (ME) Daily News*, June 8, 2012.

58. Saunders, Hachey, and Fay, "Maine's Diadromous Fish Community," 538.

59. Miller, "Two Years after Dam's Removal."

60. John Holyoke, "Nearly 1,200 Atlantic Salmon Returned to Penobscot This Year," *Bangor (ME) Daily News*, November 21, 2019.

61. Holyoke, "Breaching of Veazie Dam."

62. Aldo Leopold, *A Sand County Almanac: With Essays on Conservation from Round River* (New York: Ballantine Books, 1966), 240.

INDEX